Report 177

London, 1998

Dry weather flow in sewers

C M Ainger

R J Armstrong

D Butler

CIRIA *sharing knowledge ■ building best practice*

6 Storey's Gate, Westminster, London SW1P 3AU
TELEPHONE 0171 222 8891   FAX 0171 222 1708
EMAIL switchboard@ciria.org.uk
WEBSITE www.ciria.org.uk

Summary

Dry weather flow is all flow in a sewer that is not directly caused by rainfall. The concept is used in the design, operation and modelling of wastewater treatment works and sewer systems, and for consent setting in water quality planning. It has traditionally been defined in terms of quantity, but its quality is also important. Dry weather flow is a convenient concept, but attempts to define it more precisely lead to problems in its measurement and application.

This report reviews the dry weather flow information needs of designers, operators and modellers and compares these with current practices for gathering information. This highlights three problems: seasonal differences in infiltration, attenuation of flows within the sewer system, and difficulties in measuring pollution loads that are influenced by sediment deposition and erosion.

The review is followed by an analysis of historic dry weather flow data, which breaks new ground by assessing the effect of infiltration on the quality parameters. The results of this analysis are used to produce new guidance on *per capita* contributions to domestic flows; diurnal variation; estimation of infiltration; and estimation of crude wastewater loads when aiming for 50% solids removal in primary sedimentation. Guidance is also given on data collection and different definitions of dry weather flow appropriate for different applications.

Dry weather flow in sewers
Construction Industry Research and Information Association
CIRIA Report 177

© CIRIA 1998

ISBN 0-86017-493-X

Construction Industry Research and Information Association
6 Storey's Gate, Westminster, London SW1P 3AU
Telephone: 0171 222 8891   Facsimile: 0171 222 1708
Email: switchboard@ciria.org.uk

Keywords		
Dry weather flow, water quality, modelling		
Reader Interest	**Classification**	
Sewerage undertakers, wastewater treatment plant undertakers, and water quality planners.	AVAILABILITY	Unrestricted
	CONTENT	Original analysis
	STATUS	Committee guided
	USER	Civil engineers and water quality planners

D
627
AIN

Foreword

This report was produced as a result of CIRIA Research Project 533 *Characterisation of dry weather flow in sewers*. The report's purpose is to describe current needs and practices and to develop new guidelines for estimating the quantity and quality of dry weather flow in sewers.

The report was written under contract to CIRIA by Montgomery Watson in association with Imperial College, London. The principal authors are Mr Charles Ainger, Mr Robert Armstrong and Dr David Butler.

The project was part of CIRIA's wider programme of work on urban drainage. Other CIRIA urban drainage publications include R141 *Design of sewers to control sediment problems*, R156 *Infiltration drainage – manual of good practice* and Book 14 *Design of flood storage reservoirs*.

Following established CIRIA practice, the work was guided by a Steering Group:

Prof David Balmforth (Chairman)	Sheffield Hallam University
Neil Burns	Southern Water Services Ltd
Ian Clifforde	WRc
Kieran Downey	East of Scotland Water
Ken Mulholland	The Water Executive - Northern Ireland
Martin Osborne	Reid Crowther Consultants Ltd
Roger Saxon	Environment Agency
Mr Bob Smisson	Smisson Foundation
Graham Squibbs	North West Water Ltd
Dr Hugh Tebbutt	Independent Consultant[*]

[*] Now Professor of Water Management, Sheffield Hallam University

CIRIA's research manager for the project was Dr Judy Payne. The project was funded by Southern Water Services Limited, North West Water Limited, The Water Executive – Northern Ireland and SADWSS.

CIRIA is grateful for the help given to the project by the funders, the members of the Steering Group and all those who provided information to the research contractor.

Contents

LIST OF BOXES

LIST OF FIGURES

LIST OF TABLES

Glossary

aerobic   Conditions in which dissolved oxygen is present.

aesthetic pollution   Aspects of pollution sensed by sight or smell, such as floating solids, oil films or bankside litter.

anaerobic   Conditions in which dissolved oxygen is not present.

attenuation   The reduction in peak flow or concentration and increase in minimum flow and concentration of the *diurnal variation* in wastewater flow as it passes through the sewerage system.

bedload   That part of the sediment load that travels by rolling or sliding along the sewer invert or deposited bed, or by saltating.

biochemical oxygen demand   The amount of dissolved oxygen consumed by microbiological action when a sample is incubated in the dark at $20°C$.

catchment   An area served by a single drainage system.

chemical oxygen demand   The measure of oxygen required to oxidise all organic material in a water sample with a strong chemical, usually potassium dichromate.

combined sewer   Sewer conveying both wastewater and surface water.

combined sewer overflow   Device on a combined or partially separate sewerage system that allows flows in excess of the capacity of the system to be discharged to another sewer, to a stormwater retention tank, to a watercourse or other disposal point.

consent conditions   The conditions imposed by the regulator before permitting the discharge of a potentially polluting flow to a watercourse.

diurnal variation   The variation in flow rate or in the concentration (or mass flow) of a substance over a period of 24 hours.

domestic wastewater   Wastewater discharged from kitchens, washing machines, lavatories, bathrooms and similar facilities.

effluent   Liquid discharged from a given process.

exfiltration   The escape of wastewater from the sewerage system into the surrounding soil via cracks or malfunctioning pipe joints.

first foul flush   High concentrations of pollutants that are generated in the early part of a storm, caused by the scouring out of deposited material in the sewer system.

fixed suspended solids	Those suspended solids remaining after a sample of wastewater has been heated to 500°C.
formula A	Formula for calculating the minimum flow a sewer system overflow has to allow to pass forward to treatment before spill occurs.
foul sewer	Sewer that conveys only foul wastewater (known as a *sanitary sewer* in the USA).
gross solids	Large faecal and organic matter and other wastewater debris.
infiltration (to sewer)	Passage of water from the ground into a sewer or other conduit, via purpose-formed openings, leaking joints or defects in the conduit.
inflow	Stormwater runoff that enters a sewer indirectly.
MOSQITO	A dynamic sewer quality model.
non-volatile suspended solids	*Fixed suspended solids.*
partially separate system	Separate system in which some surface water is admitted to the sewers that convey foul water.
pollutant	Any substance conveyed in solution, suspension or as a discrete solid and discharged to a watercourse thus adversely affecting its quality.
receiving water	Watercourse, river, estuary or coastal water into which the outfall from a combined sewer overflow or wastewater treatment works discharges.
runoff	Water from precipitation which flows off a surface to reach a drain, sewer or receiving water.
sanitary solids	Fine suspended solids contained in wastewater.
sediment	Material transported in a liquid that settles or tends to settle.
self-cleansing	The ability of the flow in a sewer to transport suspended solids without a long-term build-up of deposits occurring.
separate sewer	Sewer conveying either foul wastewater or surface water, but not both.
settling velocity	Terminal velocity at which a particle falls in a still liquid (same as *fall velocity*).
sewage	*Wastewater.*
sewer	Pipeline or other conduit, normally underground, designed to convey *wastewater*, stormwater or other unwanted liquids.

sewerage system	System of sewers and ancillary works that conveys wastewater to a treatment works or other disposal point.
specific gravity	The mass of a substance divided by the mass of the same volume of water.
surface water sewer	A sewer designed to carry the runoff from paved surfaces.
suspended solids	Solids transported in suspension in the *wastewater* flow and prevented from settling by the effects of flow turbulence.
trade effluent	Wastewater discharge resulting wholly, or in part, from any industrial or commercial activity.
unionised ammonia	Ammonia dissolved in water exists mostly in the ionised form NH_4, but some is present in an unionised form NH_3. It is this latter form which is most toxic to fish.
volatile suspended solids	The largely organic fraction of *suspended solids* driven off as gas at a temperature of 500°C.
wastewater	Water discharged as a result of cleansing, culinary or industrial processes (see also *foul wastewater*).

Abbreviations

$BOD_{(5)}$	Biochemical oxygen demand (five-day)
COD	Chemical oxygen demand
CSO	Combined sewer overflow
DO	Dissolved oxygen
DWF	Dry weather flow
DWL	Dry weather load
FFT	Flow to full treatment
FOG	Fats, oils and greases
MDPF	Maximum daily peak flow
MHLG	Ministry of Housing and Local Government
MPN	Most probable number
NWC	National Water Council
PCB	Polychlorinated biphenyls
PG	Population-generated flow
P_F	Peak factor
SCFA	Short-chain fatty acid
SS	Suspended solids
TKN	Total Kjeldahl nitrogen
TOC	Total organic carbon
TSS	Total suspended solids
UPM	Urban pollution management
UWWTD	Urban Wastewater Treatment Directive
VFA	Volatile fatty acids
VSS	Volatile suspended solids
WwTW	Wastewater treatment works

1 Introduction

1.1 BACKGROUND

Information on the quantity and quality of dry weather flow in foul and combined sewers is needed in the design of treatment works and as input to water quality models used in urban pollution management. Relatively little information is available, particularly on flow quality. Guidance is needed by sewerage engineers, designers of wastewater treatment works, water quality planners and regulators on typical quantities and qualities of dry weather flow.

Dry weather domestic wastewater flows are similar around the world, but flow and quality vary considerably in detail with climatic conditions, the availability of water and its characteristics, individual domestic water consumption, diet, the presence of industrial and trade wastes, and the level of infiltration. Even within the UK, there are significant differences from location to location.

Wastewater contains a complex mixture of natural organic and inorganic materials with a small proportion of man-made substances derived from commercial and industrial practices. Relatively little is known about the detailed composition of wastewater and few specific studies have been carried out. There is also uncertainty about the quantities of pollutants entering the sewers from different sources - especially for solids and their associated pollutants - and the mechanisms of transport within the sewer. When using dry weather flows, very little account is taken of the transformations that occur as the wastewater flows through the sewer system.

The need for better guidance on dry weather flow (DWF) is urgent if sewerage undertakers are to adopt urban pollution management (UPM) procedures and meet the requirements of the Urban Waste Water Treatment Directive (UWWTD).

1.2 OBJECTIVES

The objectives of this report are:

1. To describe current practices and data requirements.

2. To review existing available data on DWF characteristics and assess its suitability for design and modelling purposes.

3. To review recent and current research on prediction of DWF and assess its potential as a practical engineering tool.

4. To identify further research requirements and associated data collection and analysis needs.

5. To provide concise guidelines on typical DWF quantity and quality for different catchment types and applications, for use by sewerage engineers, designers of wastewater treatment works and water quality planners.

1.3 SCOPE AND APPROACH

This report covers sewer flows made up of domestic wastewater, commercial and industrial discharges and infiltration flows that are not directly influenced by rainfall. It deals only with flows in combined and foul sewers. Temporal and seasonal changes in wastewater quantity and quality are considered, as is solid material that enters the sewer under dry weather flow conditions. Although infiltration is important, its estimation is not considered in detail as this has been the subject of another CIRIA study, *Control of infiltration to sewers.*

The report is written for all those interested in dry weather flow characteristics, including the designers, operators and modellers of sewerage systems and wastewater treatment works, and water quality planners and regulators.

Information on current knowledge, needs and practice has been gathered for this report by literature survey and by interviewing several water plcs, consultants and academics. Information from the literature survey and the consultation process is included in Chapters 2, 3 and 4.

Existing data on dry weather flow has been collected and analysed in order to calculate *per capita* outputs for flow and quality parameters. One of the innovations was to try to remove the effects of infiltration before analysis, to see if this resulted in more consistent figures across the range of catchments.

1.4 READER GUIDE

The term *dry weather flow* is widely used but there is no common accepted meaning. Each user has his own definition, depending on his needs. As there is usually an interface between different users – a sewerage network connects to a wastewater treatment plant and both connect to a receiving watercourse – it is important to promote better understanding to ensure effective communication. Readers are therefore urged to study sections that describe practices outside their immediate sphere of interest. The following guide is designed to help readers use the report in this way.

How does my perception of DWF compare with that of others who use the same term?	Read Chapter 2
So what is everybody interested in when they talk about DWF?	Read Chapter 3
Now I know what things others are interested in, how do they deal with them?	Read Chapter 4
Everything seems straightforward, what's the problem?	Read Chapter 5
So what's new?	Read Chapter 6
All I need to know is what numbers to use - where are they?	Read Chapter 7
Where do we go from here?	Read Chapter 8

2   Current needs

2.1   INTRODUCTION – BASIC DEFINITIONS

The dry weather flow in a sewer system is the wastewater from the domestic population, commercial properties and industry that is being transported to a wastewater treatment works (WwTW) or other disposal point. In most sewer systems the flow will also include infiltration for at least part of the year. DWF should therefore occur only in combined sewers or the foul (sanitary) part of a separate system.

The basic definition of DWF is **all flow in a sewer except that caused directly by rainfall**. The average daily DWF is given by:

$$DWF = PG + I + E \tag{2.1}$$

where

DWF	=	dry weather flow (m^3/day)
P	=	population served
G	=	average domestic wastewater contribution (m^3/*capita* day)
I	=	infiltration (m^3/day)
E	=	industrial effluent discharged in 24 hours (m^3/day)

The formula gives an average flow rate which in practice will vary through the day. E is sometimes deemed to include metered commercial discharges.

The following is an alternative definition of dry weather flow written in terms of water quality parameters of interest to designers of wastewater treatment processes:

$$DWL = PH + J \tag{2.2}$$

where

DWL	=	dry weather load of pollutant (g/day)
P	=	population served
H	=	average domestic waste contribution of pollutant (g/*capita* day)
J	=	industrial discharge of pollutant (g/day)

Values of DWL can be calculated for each pollutant of interest. It should be noted that Equation 2.2 does not include infiltration, as this is assumed not to be a source of pollutants.

It is difficult to find a universally acceptable definition of DWF because of the different ways in which it is used. The needs of users are reviewed below.

2.2 USAGE OF DWF – QUANTITY

2.2.1 Sewerage designers, operators and modellers

A designer of sewerage systems is typically interested in peak flows as given by the following equation:

$$\text{Design flow} = (\text{Peaking factor}) \times (PG + E) + I \qquad (2.3)$$

where P, G, I and E are as defined for Equation 2.1

Estimates of population and *per capita* output are usually based on some date in the future to allow for anticipated changes. Maximum values of industrial inputs and infiltration are used.

In addition to peak flows, minimum night flows are estimated to check against deposition of suspended solids and generation of hydrogen sulphide.

Sewerage operators may wish to know whether existing systems have adequate capacity. This would also be done with maximum values but using current *per capita* figures and industrial outputs.

Modellers of sewerage systems are often more interested in actual rather than theoretical maximum values, especially when they are trying to verify models against field measurements. They are also concerned with diurnal variations in DWF for simulations over periods of several hours.

In the UK combined sewer overflow settings are frequently based on Formula A (MHLG 1970):

$$Q = DWF + 1360P + 2E \qquad (2.4)$$

where

Q = flow being passed forward to treatment before spill occurs
DWF, P and E are as defined for Equation 2.1

This formula is based on DWF – including infiltration – and is used by both sewerage designers and water quality planners. It is interesting to note that the authors of the report that produced Formula A concluded that

the custom of expressing the setting [of an overflow] as a multiple of dry weather flow is basically unsatisfactory.

The report also produced Formula B and Formula C, which the authors preferred, but they felt that these formulae included terms for which insufficient data was available. With the advent of UPM procedures, acceptability of an overflow should be based on its effect on the receiving watercourse rather than Formula A.

2.2.2 WwTW designers, operators and modellers

WwTW designers, operators and modellers use DWF in a number of ways. Inlet works are generally sized to accept flows up to Formula A. Flow to full treatment (FFT) has traditionally been limited to

$$FFT = 3PG + I + 3E \qquad (2.5)$$

where

FFT = flow to full treatment
P, G, I and E are as defined for Equation 2.1

With the advent of the UWWTD, which introduces a possible requirement to treat only the diurnal cycle, there is a need to know more about actual peak flows arriving at a treatment works under dry weather conditions.

Flows between FFT and Formula A are diverted to storm tanks. The storm tanks typically hold two hours of flow at a flow rate equal to the difference between Formula A and FFT. Before the introduction of Formula A the treatment works overflow would have been set at 6DWF, so the storm tank storage would have been two hours at 3DWF, which is equivalent to six hours at DWF.

2.2.3 Water quality planners

For water quality planning and modelling, including determination of discharge consent conditions, details of wastewater flow are required. This includes estimates of summer and winter infiltration rates, and the domestic and trade components of the wastewater. If all or part of the incoming flow is pumped then measures of the flow arriving under different pumping regimes will be required. This should take account of any balancing arrangements that are proposed or in place.

They also have an interest in overflow settings and storm tank capacities because of the intermittent impacts these have on receiving watercourses.

2.3 USAGE OF DWF – QUALITY

2.3.1 Sewerage designers, operators and modellers

As well as peak flow, sewer design requires information on sediment characteristics if the sewers are to be designed as self-cleansing, although a design velocity of 0.6 m/s to 0.75 m/s at peak DWF is usually considered acceptable. In combined systems it may be surface-derived solids, such as road grit, that lead to minimum velocity requirements.

Sewer quality modellers need information on the concentrations of various quality parameters in DWF, and details of their diurnal variation. The most commonly required quality parameters are biochemical oxygen demand (BOD), chemical oxygen demand (COD), ammonia and suspended solids. Information on particle size, relative density and settling velocity is required to predict the behaviour of solids and pollutants attached to them.

Sewer quality modelling is usually carried out to estimate pollutant discharges from combined sewer overflows under storm conditions. Most of these pollutants are derived from DWF, so it is important to represent DWF accurately.

2.3.2   WwTW designers, operators and modellers

For the design of secondary treatment processes, the wastewater treatment works designer requires information on the loads of suspended solids, BOD and ammonia arriving at the works. These have typically been expressed in kg/day. If nutrient removal is to be included then information on total nitrogen and total phosphorus loads will be necessary. A wide range of values is used for calculating nitrogen and phosphorus loads.

Information on settleability, particularly of volatile suspended solids, is needed for estimating removal at primary sedimentation. This knowledge is particularly important in the light of the UWWTD requirement to remove 50% of total suspended solids and 20% of BOD at primary treatment.

Ideally, treatment works designers would like information on the levels of all other parameters that may affect the treatment process – such as temperature, pH, sulphates, heavy metals and pesticides. Increasingly the practice is to carry out field measurements at proposed treatment works sites where the sewerage system already exists.

More detailed data is required for treatment works modelling. Details are needed of how flows and quality parameters vary during the day, rather than simple estimates of total daily loads and peak flows. Treatment models can be either COD or BOD based and use either TKN or ammonia. The split between particulate and dissolved COD/BOD is required. The method for obtaining the dissolved fraction should be given in the manual for the software being used and this should be confirmed before testing is carried out.

2.3.3   Water quality planners

Water quality planners require information on the concentrations of parameters used in discharge consents – usually suspended solids, BOD and ammonia. If the discharge consents require nutrient removal the concentrations of other nitrogen compounds and phosphorus are also needed. Water quality planners are also interested in the concentrations of any substance likely to cause harm to the environment such as those on the UK Red List and the EU Black and Grey Lists.

The daily variation in flow and quality of the wastewater may also be required for modelling. Any special characteristics of the wastewater should be identified, such as large seasonal variation in inputs, significant trade effluent components, history of shock loads, major surface water inputs or sedimentation in the sewers in dry weather.

2.4   SUMMARY OF NEEDS

A summary of DWF data needs is given in Table 2.1. The needs are categorised as follows:

H   high importance
M   medium importance

Where no category is given the parameter is of low importance.

Table 2.1 *Summary of data requirements*

PARAMETER	SUBDIVISIONS	SEWERAGE			TREATMENT			WATER QUALITY	
		Design	Operations	Modelling	Design	Operations	Modelling	Planning	Modelling
Flow	Average daily flow		H	H	H	H	H		H
	Peak Daily Flow	H			H	H		H	
	Diurnal Profile, Flow			H			H		
	Diurnal Profile, Quality			H			H		
Suspended Solids	Total Suspended Solids	H		H	H	H	H	H	
	Volatile Suspended Solids				H	H	H		
	Settlability				H	H	H		
	Settling Velocity	H		H					
	Particle Size Distribution	H		H					
	Particle Density(ies)	H		H					
Oxygen Demand	Total BOD			H	H	H	H	H	H
	Dissolved BOD			H	H	H	H	H	H
	Total COD			H	M	M	H		
	Dissolved COD								
	Total Organic Carbon								
Nitrogen	Total Nitrogen			H	H	H	H		
	Ammonia			M	H	H	H	H	H
	TKN				M	M	H		
	Nitrate Nitrogen						H	H	H

...continued

Characterisation of dry weather flow in sewers

PARAMETER	SUBDIVISIONS	SEWERAGE			TREATMENT			WATER QUALITY	
		Design	Operations	Modelling	Design	Operations	Modelling	Planning	Modelling
Phosphorus	Total Phosphorus				M	M	H	M	
	Orthophosphate				M	M	H	M	
	Soluble Reactive Phosphorus								
Other Salts	Chloride	H				H			
	Sulphide/Sulphate	H				H			
Heavy Metals, Poisons, etc.					M				
Grease, Oil, etc.			H		M	H		H	
Aesthetics		H	H		M	H		H	
Pesticides & Herbicides						H		H	
Bacteriological	Total Coliforms							H	
	Faecal coliforms							H	
	Escherichia Coliforms							H	
	Faecal Streptococci								
	Pathogens							H	
Alkalinity					H	H	H		
pH					M	H	H	H	
Temperature				M	H	H	H	H	
Conductivity									H

Characterisation of dry weather flow in sewers

3   Characteristics of dry weather flow

3.1   QUANTITY PARAMETERS

3.1.1   Domestic

The quantity of domestic wastewater flow is usually expressed in terms of the volume generated per person per day, as litres *per capita* per day. This volume varies between weekdays and weekends, and seasonally. It also varies with land use and socio-economic mix, and between different regions of the UK. There may be longer-term trends because of changing lifestyles.

3.1.2   Industrial

For commercial properties, such as hospitals and hotels, flow quantity is usually expressed as daily output per bed; for schools as output *per capita*; and for office premises and warehouses as output per unit area of floor space. Most industrial outputs are subject to trade waste consents that lay down maximum volumes of discharge per day and per three-month period. These give useful estimates of maximum flows to designers and operators but can lead to confusion during analysis of actual flow measurements, as outputs are not always at their maximum. Sometimes the total commercial and industrial flows are divided by the domestic *per capita* output to give an effective additional population. This is then added to the resident population to give a population equivalent.

3.1.3   Infiltration

Sewerage systems are subject to the intrusion of groundwater (infiltration) and possible loss of wastewater into the surrounding ground (exfiltration) through cracks and fissures, pipe joints and couplings, and manholes. The foul part of a separate sewerage system can also suffer from ingress of rainwater through illegal or misconnected surface water yard gullies, roof downpipes or through manhole covers. This response to storm inputs is referred to as inflow. Infiltration is considered a part of DWF but inflow is not.

Some infiltration contributes to DWF most of the time. Understanding and quantifying it is important because the presence of excessive amounts may cause one or more of the following problems (Ledbury, 1982; Fiddes and Simmonds, 1981):

- increased sediment entry resulting in higher maintenance requirements and possible surface subsidence due to erosion of the surrounding soil

- effective capacity of the sewer is reduced, resulting in possible surcharge conditions and (exceptionally) flooding

- discharge of diluted raw or poorly treated wastewater at overflows during periods of high groundwater table or rainfall

- overloading of pumping stations and wastewater treatment works.

Infiltration also leads to additional pumping and treatment costs.

3.2   QUALITY PARAMETERS

3.2.1   Range and units

A full list of the quality parameters that could be considered in dry weather flow would be extensive. To simplify matters, this report concentrates on those pollutants found in dry weather flow from domestic premises. Industrial discharges are subject to consents that are often set in terms of COD and suspended solids. However, in the UK the bodies responsible for operating wastewater treatment works review the full spectrum of chemicals likely to be discharged to sewers by each industry, and determine their acceptability in terms of treatability or effect on treatment processes. Substances outside of these consents should be regarded as illegal discharges and should not therefore be considered in a characterisation process. These illegal discharges can sometimes be caused by accidental spillage.

From the point of view of treatment works designers and operators, the major quality parameters are usually expressed in terms of mass (kg) per day. For sewer quality modelling, WwTW modelling and for receiving water quality planning, quality parameters are more commonly expressed as concentrations. The actual measurement of quality parameters is always in terms of concentration and conversion to mass flow requires a corresponding measurement of flow rate.

For the purposes of characterising DWF, mass flow rate is the better parameter because concentration varies with infiltration. Caution has to be applied when calculating mass flow – it is **not** the average concentration from a number of samples taken over a day multiplied by the average daily flow. This is because both concentration and flow rates vary over the period of a day, and peak concentrations usually occur at times of peak flow.

The quality parameters that need to be considered are described below.

3.2.2   Oxygen demand

Oxygen demand is one of the most commonly used measures of DWF quality. It is important as a measure of the depletion of dissolved oxygen in receiving waters, which damages aquatic life. Limits on oxygen demand are usually set for treatment works effluents, so it is important for treatment works designers and operators to know the oxygen demand of the incoming flow and how much has to be removed. Oxygen-demanding substances can also enter watercourses from combined sewer overflows during wet weather. DWF has a high oxygen demand and knowledge of this is needed to predict the quantity in overflow spills.

That part of the oxygen demand which is readily biodegradable is measured by the biochemical oxygen demand (BOD) test. The standard BOD test, BOD_5, takes five days. Other measures of BOD, such as the 20-day BOD, are sometimes required for treatment models. BOD is a subset of the total amount of substances that can be oxidised by chemical means, the chemical oxygen demand (COD). COD is quicker and easier to measure than BOD so efforts have been made to relate BOD to COD loads on a site-specific basis. No universal relationship has been found between the two parameters.

COD tests do not differentiate between biodegradable and inert organic matter or between readily and slowly biodegradable fractions. Wastewater characterisation by COD fractionation is therefore regarded as an indispensable step for the reliable modelling of biological treatment processes (Orhon, 1994). This is particularly so if plant optimisation is being carried out or if nutrient removal stages are proposed.

Total influent COD has two major components: total non-biodegradable or inert COD and total biodegradable COD. It can be further divided as shown in Figure 3.1. The soluble fraction of readily biodegradable organic matter (rapidly hydrolysable) constitutes 10-15% of raw wastewater total COD. The hydrolysis under aerobic conditions is rapid and will be complete within a few hours. Hence the extent of the sewer network will alter the size of this fraction considerably. Organic compounds that can be directly metabolised are limited to small molecules of volatile fatty acids (VFA, also known as short-chain fatty acids (SCFA)). Carbohydrates, alcohols, peptones, and amino acids make up most of this fraction. VFA, ethanol, and glucose can account for between 50% and 70% of the readily available COD in DWF.

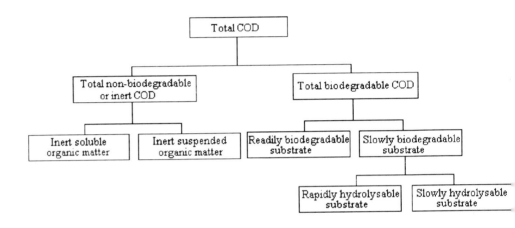

Figure 3.1 *Classification of COD fractions*

An alternative to COD is a total organic carbon (TOC) test. This is sometimes used for reasons of speed and reliability. There is no universal relationship between TOC and BOD, but there is a theoretical relationship between TOC and COD:

$$C(12) + O_2 (32) = CO_2 (44)$$

hence $TOC \times 2.67 = COD$

Oxygen-demanding substances can be fully dissolved in the DWF, of colloidal or fine particulate nature, or attached to other matter suspended in the flow. A knowledge of the relative amounts of particulate and dissolved oxygen demand is required for some purposes. Methods used for measuring the dissolved portion involve removing the suspended matter by settlement or filtration. Different removal methods give different results so it is important to be clear on the method used.

3.2.3 Solids

Wastewater contains solid material generated by domestic appliances. Most of these sanitary solids are finely divided and carried in suspension in the bulk sewer flow. There are some larger *gross solids* (greater than 6 mm in two dimensions, also known as aesthetic, refractory or intractable solids) that are introduced into the system by water closets. Although not comprehensively defined, these include some faecal stools, some toilet paper and most sanitary refuse. Meeds and Balmforth (1995) have suggested the following categories (the list includes solids that enter the sewer in surface runoff):

condoms
disposable nappies
faeces

fine tissue paper
leaves
miscellaneous (paper origin)
miscellaneous (fat origin)
paper towels
plastic
sanitary towels
sweet wrappers
tampons.

Less common items are toothbrushes, razor blades, syringes, shoes, rags or even puppies or kittens (Burchmore and Green, 1993; Friedler *et al.*, 1996; National Bag It and Bin It Campaign, 1995). It is the gross solids that cause aesthetic pollution of receiving waters from combined sewer overflows.

Fine solids are important in sewer system design, which should ensure that they are kept in suspension and transported to the outlet. To determine design velocities one or more of the following parameters needs to be known: particle size, size distribution, relative density and settling velocity. These parameters govern the rate at which sewer sediments build up during dry weather. Deposited sediments can contribute to *first foul flush* effects during storm events. Treatment works designers use *settleability* as a measure of how quickly particles will settle and therefore how many will be removed by primary sedimentation.

Butler *et al.* (1996) have suggested three classes of solids: sanitary solids, stormwater solids (both transported mainly in suspension and measured by the total suspended solids (TSS) test) and grit (carried mainly as bedload). It is the sanitary solids that are most associated with DWF, but quantities of each of these solids may be transported under dry weather conditions. During low flow periods, solids may deposit on the pipe invert or sediment bed but will be re-eroded as flow builds up once again. TSS may be further sub-divided into volatile and non-volatile. Volatile suspended solids (VSS) gives an indication of organic content.

3.2.4   Nitrogen

Nitrogen is an important element since biological reactions can only proceed if it is present in sufficient quantities. Nitrogen exists in four main forms: organic, ammonia, nitrite and nitrate; but is expressed in a variety of ways as shown in Box 3.1.

Box 3.1 *Forms of nitrogen*

Total Nitrogen	Organic nitrogen, ammonia, nitrite, and nitrate.
Organic Nitrogen	Molecular nitrogen bound to proteins, amino acids, urea and various chemicals and solvents.
Ammonia (NH_4)	Ammonia nitrogen is produced at the first stage of decomposition of organic nitrogen (either as ammonium salts or free ammonia).
Total Kjeldahl Nitrogen (TKN)	Organic and ammoniacal nitrogen.
Nitrite (NO_2) and Nitrate (NO_3)	Nitrate is nitrogen in its most highly oxidised form. Nitrite is an intermediate oxidation state not normally present in large amounts.

One of the most commonly used measures of nitrogen is total ammonia and discharge consents for WwTW often set a limit on the concentration of total ammonia. Despite this, it is unionised ammonia above a certain concentration threshold that is most toxic to fish. The ratio of unionised ammonia to total ammonia varies with pH and temperature of the receiving water and its toxicity depends on the dissolved oxygen concentration (Foundation for Water Research, 1994a). Some modellers of treatment works processes prefer to work in terms of Total Kjeldahl Nitrogen (TKN) in order to carry out mass balance calculations. When using this measure of total nitrogen it is possible to keep track of conversions between ammonia, nitrate and nitrite.

Nitrogen in the form of nitrate is a nutrient and can promote weed and algal growth (eutrophication) in receiving waters. Apart from the hydraulic and aesthetic effects, respiration of the algae and weeds can reduce dissolved oxygen levels. Nutrient removal is sometimes required by discharge consents.

3.2.5 Phosphorus

Phosphorus, like nitrogen, is a nutrient and in fresh waters is most often rate limiting for algal growth. If the small amount of phosphorus that is present in wastewater is removed, then excessive algal growth will be prevented whatever the nitrogen concentration.

The major sources of phosphorus in domestic wastewater are excreta (50-60%) and polyphosphate builders in synthetic detergents (35-50%). Phosphorus concentrations have diminished dramatically in areas where legislation has imposed significant reductions in the amounts of phosphorus used by manufacturers of synthetic detergents. Phosphorus can be expressed as total, organic or inorganic phosphorus (see Box 3.2). The usual forms found in solution include orthophosphates, polyphosphates and organic phosphate. Orthophosphates are directly available for biological metabolism, but polyphosphates must undergo (rather slow) hydrolysis to become available. Organically bound phosphate is usually of minor importance.

Box 3.2 *Forms of phosphorus*

Total Phosphorus	Total phosphorus exists in organic and inorganic forms and is associated with eutrophication of receiving waters.
Organic Phosphorus	Organic phosphorus is bound in organic matter.
Inorganic Phosphorus	Inorganic forms of phosphorus exist as orthophosphate, metaphosphate or polyphosphate.

3.2.6 Sulphates

Domestic wastewater contains sulphur compounds, and effluents from the meat, leather, brewing and paper industries can also contain very high concentrations. Organic sulphur compounds are present in excreta and industrial effluents. The sulphate present in wastewater is derived principally from the municipal water supply but it can also be derived from saline groundwater infiltration (Thistlethwayte, 1972).

Under anaerobic conditions, organic sulphur compounds and sulphates are reduced to form sulphides, mercaptans and other compounds. The principal product, hydrogen sulphide, is formed mainly in the slime that grows on the wall of sewers. The most favourable conditions for its production are small-diameter pipes filled with anaerobic wastewater for a long period, at a high temperature.

Hydrogen sulphide can contaminate the atmosphere in manholes, gravity sewers, and wet wells of pumping stations. It is a flammable, toxic gas that can cause serious odour nuisance. It is acutely toxic to aquatic organisms and could be a factor in fish kills near combined sewer overflows as it reacts immediately with any dissolve oxygen present, thereby rapidly reducing DO concentrations. Hydrogen sulphide in damp conditions can damage concrete, electrical equipment, step-irons, and ladders (Boon, 1992) through chemical corrosion. While this does happen in the UK the problem is not as widespread as it is in hot climates. Methodologies for estimating the rate of generation of hydrogen sulphide have been developed (Pomeroy, 1976).

Sulphides in concentrations of 25 mg/l or more completely inhibit biological growth in non-acclimatised activated sludge processes. After a few days' adaptation the tolerance increases to 100 mg/l (Degremont, 1973). Hydrogen sulphide also causes problems at treatment works, where it reduces the settleability of sludge.

3.2.7 Alkalinity

Alkalinity results from the presence of bicarbonates, carbonates and hydroxide compounds of calcium, magnesium, sodium and potassium; and is expressed as mg/l $CaCO_3$. Wastewater alkalinity is derived from water supply, groundwater and domestic chemicals. The alkalinity in wastewater helps to neutralise any added acids and is particularly important where biological nutrient removal is practised or proposed.

3.2.8 Microbiological parameters

Despite the importance of coliform bacteria as indicators of the presence of pathogenic micro-organisms, little is known about their numbers or behaviour in sewerage systems. It is supposed that most coliform bacteria are introduced into the system via the WC, but they are also present, in small amounts, in bath and laundry wastewaters (Siegrist *et al.*, 1976).

Figure 3.2 shows the diurnal pattern of total coliforms with flow in sewer samples taken in Dundee.

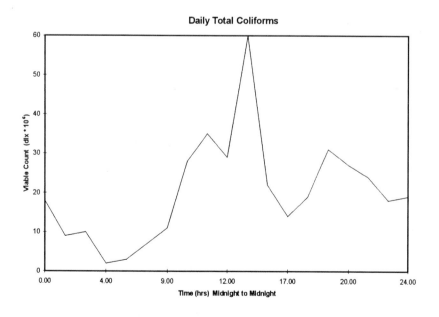

Figure 3.2 *Daily total coliforms (Jefferies* et al, *1990)*

According to Ashley and Dabrowski (1995), there is a relationship between the number of faecal coliform bacteria in flows of wastewater and other physico-chemical water quality parameters. However, this correlation is dependent on catchment characteristics and the time of year at which the samples are taken. Bacterial growth occurs in sewer slimes, and subsequent scouring-off increases concentrations in the flow.

Payne and Moys (1989) report that concentrations of bacteria are influenced by season. During dry weather and in sediment deposits, the ratio of faecal to total coliforms is lower during the summer than in the winter.

3.2.9   Temperature

Wastewater is warm (18°C in summer and 10°C in winter) because of warm wastes from residential and industrial areas. Extremes of temperature can be caused by industrial discharges. Temperature is important because it affects the rate of chemical and biological processes in the sewer and at the treatment plant. It is also important in terms of its impact on receiving water quality because it influences dissolved oxygen saturation levels.

3.2.10   pH

The pH is a measure of the acidity of wastewater. Domestic wastewater is normally slightly alkaline (pH 7.2 in hard water areas) and is heavily influenced by the pH of the local municipal water. Higher pH values are usually associated with industrial wastes.

The pH value is important because the range that supports biological life is limited to between pH 6 and pH 9. Wastewater with adverse pH is difficult to treat by biological means without acclimatisation. Chemical reactions are also very much linked to pH values.

3.2.11   Chloride

The chloride concentration of wastewater is mostly due to the presence of common salt, which is a normal constituent of urine. Thus, the chloride content is more or less constant, proportional to population, except when influenced by trade wastes (Escritt, 1959). Chloride can also be introduced into sewers in coastal areas by saline intrusion, particularly at high tide.

The effects of excessive chloride content on treatment processes include the following:

- increased effluent suspended solids and increased effluent turbidity
- decrease in organic removal efficiency
- loss of reactor volatile solids
- significant reduction in the levels and type of protozoa in the activated sludge
- inhibition of phosphorus removal
- inhibition of nitrification.

3.2.12   Fats, oils and grease

Fats is a general term often used to describe the whole range of fats, oils and greases (FOG) discharged into the sewer. They are among the more stable organic compounds and are not easily degraded biologically. The major sources of fats are food preparation (butter, margarine, vegetable fats and oil, meats, cereals, nuts and some fruit) and, to a lesser extent, excreta. Fats are only sparingly soluble in water and are converted by hydrolysis to yield fatty acids.

Fats cause problems in sewer systems because they can solidify and cause blockages. To mitigate such effects from commercial premises, purpose-built fat and grease traps can be installed. These must be emptied regularly by a registered waste contractor.

The increased use of automatic dishwashers, which operate at high temperatures and use powerful detergents, allows fat to travel further down the drainage system where it congeals in public sewers rather than private drains. These effects may also cause grease traps to be by-passed.

Restaurants are not subject to trade effluent controls but could be prosecuted under the Environmental Protection Act 1990, the Water Industry Act 1991, or the Health and Safety at Work Act 1974 – if it can be proved that their discharges are creating a nuisance. At least one water service company in the UK issues guidelines on waste disposal to catering premises.

3.2.13 Other substances

Many other substances can be present in DWF including heavy metals, herbicides, pesticides, polychlorinated biphenyls (PCBs) and phthalates. Most of these are toxic in all but the smallest trace quantities. In some cases their discharge to receiving waters is strictly limited to safeguard aquatic life. Some substances interfere with the biological processes at treatment works by killing the microbes that carry out the treatment. Other substances can be toxic to humans.

Most of these substances are from industrial sources. However, a study by Comber and Gunn (1996) has identified a number of sources of heavy metals in domestic wastewater. These include copper and lead from pipework, zinc from galvanised cisterns and other heavy metals from household products, baths, dishwashers and washing machines.

A number of substances, usually of industrial origin, have been considered so toxic, persistent or bio-accumulative that priority has been given to eliminating them from the environment. These are contained in the European Framework Directive (1976), known as the European Black List. A subset of these substances is listed in UK legislation on Environmental Protection (Prescribed Processes and Substances) Regulations 1991 and Trade Effluents (Prescribed Processes and Substances) Regulations 1989. This subset is called the UK Red List. The Framework Directive also contains a Grey List of other harmful substances. Discharge of such substances to sewers is either prohibited or their concentrations are severely restricted in the UK. DWF should not include significant quantities of these substances.

3.3 DIURNAL VARIATION

Both flow rate and quality parameters vary diurnally. A diurnal pattern from domestic properties is illustrated in Figure 3.3.

Diurnal patterns vary with the day of the week and with the season, and there may also be long-term trends. In recent years seasonal trends have become somewhat exaggerated, with water demands much higher in summer. Diurnal variations may vary within a sewerage system because of attenuation effects. Diurnal profiles can also be affected by varying levels of infiltration.

Because of diurnal variation, treatment works and sewer systems are designed to handle peak as well as minimum flows. Traditionally in the UK treatment works have been designed for a flow to full treatment defined by Equation 2.5.

Domestic daily water usage patterns

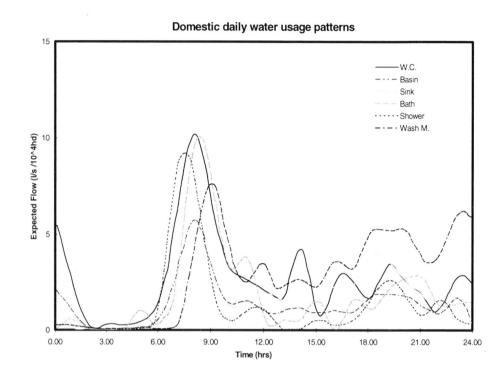

Domestic daily water usage patterns

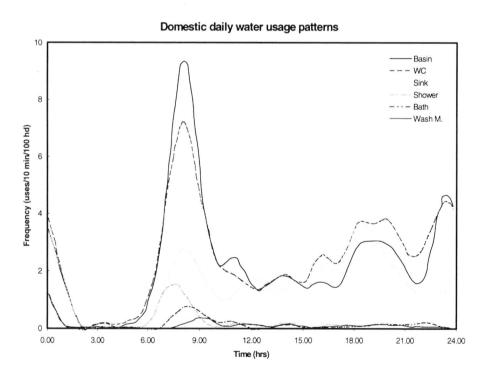

Figure 3.3 *Domestic daily water usage patterns* (*after Butler and Graham, 1995*)

The UK application of the UWWTD suggests that works may be designed to treat the diurnal cycle. This has led to the concept of maximum daily peak flow (MDPF) which is defined as the maximum flow arriving on any dry day. For many medium to large treatment works the MDPF may be significantly lower than the flow calculated from Equation 2.5. Using the definition of a dry day as *a day with no rain following a 24-hour period with no rain,* Luck and Ashworth (1996) plotted the daily dry weather flows arriving at a particular site in August 1995. These plots are shown in Figure 3.4. The plots exhibit an unusual level of variation as the site was subject to tidal influences. They are included here as a reminder that tidal influences can have a significant effect on the characterisation of DWF in coastal areas. It is expected that plots for inland sites would also show daily variations but perhaps with not such a wide range.

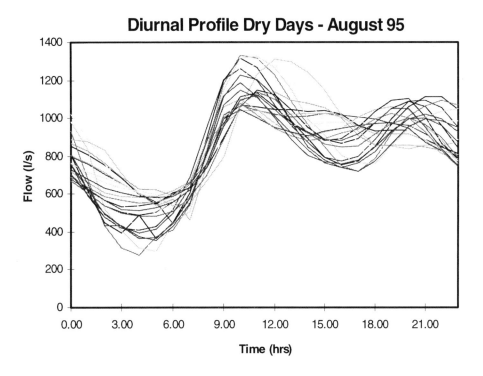

Figure 3.4 *Variation of daily peak flow (after Luck and Ashworth, 1996)*

4 Current knowledge and practice

4.1 ESTIMATION OF DOMESTIC DWF

4.1.1 Average flows and loads

In the domestic environment, water is used in three main areas. Approximately a third is used for WC flushing, a third for personal washing, and the final third for other uses such as washing up, laundry and food preparation (see Table 4.1). Changes in water use patterns and appliance ownership may alter these proportions. Implications for DWF are discussed later in Section 4.1.4 *Water consumption trends.*

Table 4.1 *Quantity of water consumed for various household purposes*

Component	Water consumed (litres/*capita* day)	Water consumed (%)
WC flushing	52	35
Washing/bathing	37	25
Food preparation/drinking	22	15
Laundry	15	10
Car washing/garden use	7	5
Other	15	10
TOTAL	148	100

Most of this water is returned for treatment via the sewerage system. That which does not return is either consumed or used externally for garden watering and car washing. Clearly, there is a strong link between water consumption and wastewater generation. It is estimated in the UK that about 95% of water used appears as DWF in the sewer network (DoE, 1992).

The WC discharge volume is typically 9 litres, although it should be noted that, since January 1993, water by-laws require that all new WCs installed to have a maximum flush volume of 7.5 litres. Average shower volumes are 50% of bath volumes, or less. Discharge volumes for washing machines have reduced since the 1970s when discharge volume was estimated by Rump (1978) as 115 litres. More recently, machines have reduced consumption to 90 litres, with water-efficient machines using as little as 60 litres. Dishwashers have also experienced a reduction, from 50 litres in the 1970s to water-conserving models now discharging just 20 litres. Increases of ownership and improvements in performance have resulted in substantial increases in the frequency of use of washing machines, automatic dishwashers and showers.

Butler (1991,1993) has shown how appliance usage frequency and volume influences the make-up of domestic DWF inputs. Figure 4.1 shows the combination of appliances that is responsible for the characteristic diurnal pattern.

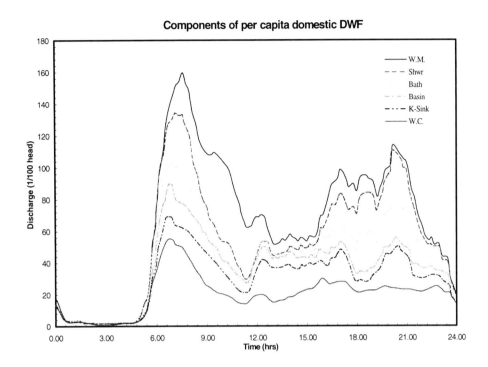

Figure 4.1 *Components of* per capita *domestic DWF (after Butler and Gatt, 1996)*

A review of UK texts shows a wide range of reported figures for *per capita* DWFs. Imhoff (1971) suggests the range of flows is 140 to 300 litres *per capita* per day, with a mean of 180 litres. Bartlett (1979) suggests water usage is rarely less than 140 litres *per capita* per day. Escritt (1984) quotes measurements made on a new housing estate with no infiltration which gave a *per capita* value of 122 litres per day. He estimated *per capita* values at 68 litres in villages, 113 litres in medium-sized towns, and 142 litres in large towns.

The most recent data set for England shows a mean water consumption figure of 145 litres *per capita* per day (Russac *et al.*, 1991; Edwards and Martin, 1995) and hence, using the 95% translation quoted earlier, a domestic wastewater flow of 138 litres *per capita* per day. However, there is a considerable range of values between individuals. Higher figures (25%) have been reported in Scotland (Gray, 1989), primarily due to the widespread use of a 13.6-litre WC cistern. Studies commissioned by the water industry have shown that *per capita* consumption is rising by about 1% a year in the UK.

Current design guidance on DWF estimation is included in the British Standard on Sewerage (BS8005:1987). The following advice is given:

Typical discharge figures for similar developments to those under consideration may... be used. In the absence of such data a figure of 220 l (200 l + 10% infiltration) may be assumed which when multiplied by the population gives the average flow or dry weather flow.

In many broadly residential catchments there will be buildings, other than domestic dwellings, such as schools and hospitals. Typical values of wastewater production from commercial sources are given in Section 4.2. However, for boarding schools, hotels, hospitals and similar residential establishments, the flow could be taken as that from normal domestic dwellings.

4.1.2   Peak flows

Peak DWFs in foul sewers can be estimated by several methods, as described below.

Building drainage and small sewerage schemes are designed using the **discharge unit** method. The most commonly used version of this method is that recommended by the British Standard on Building Drainage (BS8301:1985). The method uses the principles of probability theory, and discharge units are assigned to individual appliances to reflect their relative load-producing effect. Addition of the relevant discharge units enables peak flow rates from groups of appliances to be estimated. A more detailed explanation of this method is given by Wise and Swaffield (1995).

For larger sewerage schemes, a different approach is taken. As previously described, the flow of wastewater from a residential area has a distinct diurnal pattern – low flows occur at night, with peak flows occurring during the morning and evening. It is therefore possible to define an average DWF and describe peak flow as a multiple of this average flow. BS8005:1987 states:

Foul sewers are frequently designed to carry four to six times DWF, the larger figure relating to sub-catchments and the smaller to trunk sewers. This takes account of diurnal peaks and the daily and seasonal fluctuations in water consumption together with an allowance for extraneous flows such as infiltration.

In a similar way, *Sewers for Adoption* (Water Services Association, 1995) suggests that peak flow for foul sewers serving residential developments is 4000 litres/unit dwelling/day. This is equivalent to three persons per property, each contributing 200 litres/day, with a peak of six times DWF and 10% infiltration.

This derivation does not match current knowledge on *per capita* contributions (see Section 7.2).

A third approach is based on US practice. Average flows are estimated *per capita* and peak flows are determined by the application of peak factors. North American engineers have established that peak flow rates generally decrease from inlet to outfall of the sewerage network and have therefore related the magnitude of the peak factor to position in the network. This is accomplished by determining the population served or the average flow rate at a particular point. According to the ASCE/WPCF *Manual of Practice for Gravity Sanitary Sewer Design and Construction* (1982) peak flows can be determined quantitatively using peak factors related to population served (see Figure 4.2). This relationship with population has also been described algebraically with equations typically of the form:

$$P_F = \frac{5}{P^a} \qquad\qquad\qquad\qquad (4.1)$$

where

P_F =   peak factor
P =   population drained in 1000s
$0.15 < a < 0.2$.

In combined sewers, peak DWF is not a significant issue for sewer capacity, as dry weather flow represents a small portion of design storm flows.

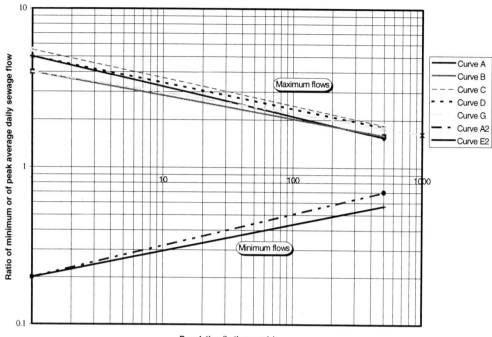

Figure 4.2 *Ratio of extreme flows to average daily flow (after ASCE/WPCF, 1982)*

Curve A source: Babbit, H.E., "Sewerage and Sewage Treatment." 7th edn., John Wiley & Sons, Inc., New York (1953)

Curve A2 source: Babbit, H.E., and Bauman, E.R., "Sewerage and Sewage Treatment." 8th edn., John Wiley & Sons, Inc., New York (1958)

Curve B source: Harman, W.G., "Forecasting Sewage at Toledo under Dry-Weather Conditions." *Eng. News-Rec,* 50, 1233 (1918)

Curve C source: Youngstown, Ohio, report

Curve D source: Maryland State Department of Health curve prepared in 1914. In "Handbook of Applied Hydraulics." 2nd edn., McGraw-Hill Book Co., New York (1952)

Curve E source: Gifft, H.M., "Estimating Variations in Domestic Sewage Flows." *Waterworks and Sewerage.* 92, 175 (1945)

Curve F source: "Manual of Military Construction." Corps of Engineers, United States Army. Washington D.C.

Curve G source: Fair, G.M., and Geyer, J.C., "Water Supply and Waste-Disposal." 1st edn., John Wiley & Sons, Inc., New York (1954)

Curve A2, $\dfrac{5}{pe.167}$ Curve B, $\dfrac{14}{4+\sqrt{p}}+1$ Curve G, $\dfrac{18+\sqrt{p}}{4+\sqrt{p}}$ where p equals population in thousands

4.1.3 Variability

Pioneering work on DWF measurement in sewers was carried out in South Africa more than 30 years ago. Shaw (1963) suggested construction of what he called **contributor hydrographs**. These were obtained by gauging relatively small catchment areas of uniform composition. He also found that the day-to-day variation in flow for any particular catchment was significant, and to develop a hydrograph suitable for use he suggested the following expression:

$$Q = Q_o + Z s \qquad\qquad (4.2)$$

where

Q	=	flow rate
Q_o	=	mean value of recorded flows
Z	=	confidence coefficient
s	=	standard deviation of recorded flows

After carrying out a statistical analysis using various integral values of the coefficient Z he suggested a value of $Z = 2$ would be suitable for design purposes. If the data is normally distributed this would be equivalent to assuming a confidence limit of approximately 95%. This work clearly showed the stochastic nature of DWF patterns.

Further measurements along the same lines were made in Johannesburg by Stephenson and Hine (1985, 1986) and similar results were obtained. To facilitate computer simulation the contributor hydrographs were fitted to a series of sine waves of different amplitude, located correctly in time.

In the USA, a recent study has been reported by Gaines (1989) in the city of Denver, Colorado. This consisted of some 3500 measurements of flow in the range 0.28 litres/s to 4248 litres/s. Rates of flow greater than the average peak flow rate were plotted and found to approximate to the normal probability distribution. Gaines also produced algebraic equations to fit these curves:

$$P_F = aQ^{-b} \tag{4.3}$$

where

P_F =   peak factor
Q =   average flow
$1.358 < a < 5.155$
$0.025 < b < 0.064$

It is interesting that both Gaines and Shaw noticed the normal distribution of measured sewer flows.

The most recent UK research in this area is by Butler and Graham (1995), who developed a model to predict the spatial and temporal variation of domestic DWF in sewer networks. The stochastic nature of DWF impacts and DWF patterns was recognised and the binomial distribution was used to model inflow from individual appliances. The necessary input data was provided by a small-scale domestic appliance usage survey and the model was verified on a small combined sewer network using 25 days of DWF data. The flow data was again found to be normally distributed at any particular time of day, producing similar but individual DWF diurnal plots. The accuracy of the predicted mean daily and peak flows fell within ± 10% of measured values and the overall fit of the data throughout the day was found to be good (see Figure 4.3). Flows at the ± 95% confidence level were also well modelled.

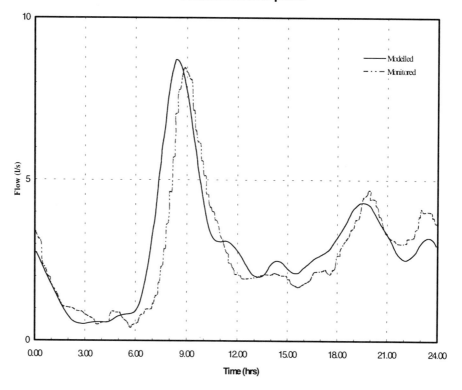

Figure 4.3 *Prediction of DWF profile (after Butler and Graham, 1995)*

4.1.4   Water consumption trends

It is difficult to predict future trends in water consumption. This section describes possible changes that might have an effect on *per capita* DWF.

Binnie and Herrington (1992) have predicted that climate change in the UK will lead to an increase in domestic water consumption of 10% by 2031, due to increased garden watering and personal showering. Perhaps 5% might be returned as DWF. This could be reduced by water metering: it has been demonstrated in metering trials on the Isle of Wight that a 10% reduction in demand can be expected if metering is widely introduced.

Demographic changes will occur and these will bring about commensurate changes in water demand. In particular, any increase or decrease in household occupancy levels is likely to cause a commensurate increase or decrease in water consumption, although there is an economy of scale as increase in occupancy effectively reduces the *per capita* demand. In the UK there is a downward trend in occupancy level. Changes in population structure, such as a high proportion of retired people, will also result in changes in the amount of water consumed and hence in the amount discharged for treatment (Russac *et al.*, 1991).

The use of cheap-rate night-time electricity for dishwashers and washing machines may increase night time DWF. Water supply companies themselves balance flows by pumping at night to take advantage of cheap-rate electricity, and storing water during the day. Again, water metering will affect these patterns if water is cheaper at night.

A major feature of demand-side management, an area of acute concern (CIWEM,1996), is the introduction of water saving appliances (minimum flow plumbing fixtures such as low-flow taps, low-flush WCs, compost and chemical toilets) and water-efficient luxury appliances (such as ecolabelled washing machines and dishwashers). It is likely that these will be steadily introduced on the basis that they provide comparable service, are compatible with existing fixtures and are cost-effective (NRA, 1995).

Reduction in water use has been brought about in the past by water rationing with public co-operation. An important element for the future will be education of the public on the need for water conservation and the measures needed to implement it. Changes in the use of water-using appliances have already occurred, with increased ownership of washing machines, dishwashers and showers. These changes will continue to affect DWF.

4.1.5   Domestic wastewater quality

The major quality parameters and their importance are described in Section 3.2. Table 4.2 draws together typical pollutant levels in UK domestic DWF, including physical, chemical and microbiological characteristics. Values given are means, in some cases followed by a representative range (in brackets).

The data is expressed as load *per capita* (which has been argued to be the more fundamental value) and as concentration. The variation in concentrations (denoted by brackets) is caused mainly by dilution due to varying degrees of infiltration. Values in the table have been compiled from many sources, some in terms of *per capita* load and some as concentration. To try to ensure consistency within the table, the values presented are linked by a standard *per capita* consumption. A value of 200 litres *per capita* per day is used.

Values for total BOD_5 and suspended solids are useful, but more detailed knowledge can be useful for optimising treatment plant design. Detailed studies in France (Chebbo *et al.*, 1990) have shown that suspended solids in sewers under DWF consist predominantly (25% by mass) of fine particles ($30<d_{50}<38$ μm). Work in Brussels (Verbanck *et al.*, 1990) supports this assertion and also shows that approximately one third is mineral in content. Fair *et al.* (1966) state that specific gravities (SG) of sanitary solids range from less than 1.0 to 1.2 on a dry basis, but may be as low as 1.001 on a wet basis. Chebbo *et al.* (1990) found the SG to be 1.6 for particles less than 100 μm and 1.4 otherwise.

Approximately 60% of suspended solids (200 mg/l) can normally be removed by sedimentation given adequate time (one to two hours). However, settleability depends on the range of particle sizes, their densities and hence particle settling velocity. These can vary from catchment to catchment (see Figure 4.4). It has also been observed that pollutant concentration is correlated with particle settling velocity fraction. Becker *et al.* (1996) for example, showed that most of the particulate COD and phosphorus in wastewater is associated with solids with settling velocities in the range 0.04 to 0.9 m/s.

Table 4.2 *Typical major pollutant characteristics in domestic DWF (compiled from a number of published sources)*

Parameter type	Parameter	Load g/*capita* day	Concentration mg/l (based on 200 litres/*capita* day)
Physical	Suspended solids		
	Volatile	48	240
	Fixed	12	60
	Total	60	300 (180-450)
	Gross solids		
	Sanitary refuse	0.15*	
	Toilet paper	7	
	Temperature		18 (15-20)°C: summer
			10°C: winter
Chemical	BOD_5		
	Soluble	20	100
	Particulate	40	200
	Total	60	300 (200-400)
	COD		
	Soluble	35	175
	Particulate	75	375
	Total	110	550 (350-750)
	TOC	40	200 (100-300)
	Nitrogen		
	Organic N	4	20
	Ammonia	8	40
	Nitrites		0
	Nitrate		<1
	Total	12	60 (30-85)
	Phosphorus		
	Organic	1	5
	Inorganic	2	10
	Total	3	15
	pH		7.2 (6.7-7.5): hard water
			7.8 (7.6-8.2): soft water
	Chlorides	20	100
	Alkalinity	20	100 dependent on water supply
	FOGs	20	100
Microbiological	Total coliforms		10^7-10^8 MPN/100 ml
	Faecal coliforms		10^6-10^7 MPN/100 ml
	Viruses		10^2-10^3 infectious units/100 ml

*   items/*capita* day

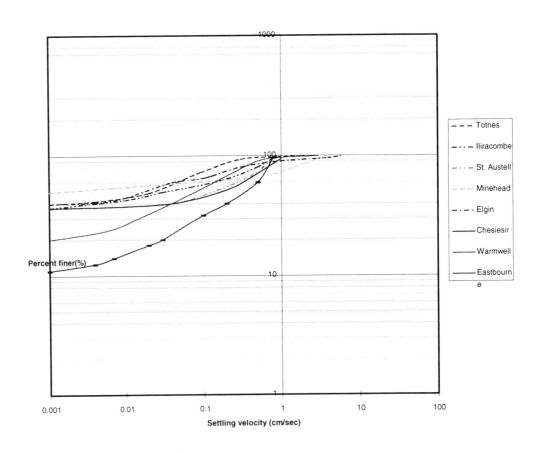

Figure 4.4 *Settling characteristics of wastewater from several catchments (after Andoh and Smisson, 1996)*

4.2   ESTIMATION OF COMMERCIAL AND INDUSTRIAL DWF

Commercial usage of water includes wastewater from businesses such as shops, offices, sports complexes and light industrial units. Typical commercial establishments include restaurants, laundries, pubs and hotels.

Demand is generated by drinking, washing and sanitary facilities, but patterns of use are inevitably different to those generated by domestic usage. For example, Table 4.3 shows how toilet usage is an even more dominant component of water use (63%) than in the domestic environment. This factor may give rise to significant 'export' of BOD by commuters from residential to commercial areas.

Estimates of flow are sometimes based on the number of employees (allowing about 20% of the domestic consumption per employee) and sometimes on the floor area of the commercial district (Barnes *et al.*, 1981). Table 4.4 gives examples of annual wastewater volume produced by a variety of commercial sources.

CIRIA Report 177

Table 4.3 *Proportion of water consumed for various office purposes (Gray, 1989)*

Component	Water consumed (%)
WC flushing	43
Urinal flushing	20
Washing	27
Canteen use	9
Cleaning	1
TOTAL	100

Table 4.4 *Annual volume of wastewater produced from various commercial sources (Henze et al., 1995)*

Category	Volume of wastewater (m³/yr)	Per
Places of work	15-20	Employee
Camping sites	25-30	Person per day
Cottages	40-60	Cottage
Military installations	50-60	Permanent resident
	15-20	Employee
Hospitals	150-250	Bed
Nursing homes, sanatoriums	100-150	Bed
Hotels, boarding houses	60-100	Bed
Restaurants etc.	100-150	Employee
Swimming baths	50-60	Visitor per day

The component of DWF generated by industrial processes can be an important one in specific situations, but it is more difficult to characterise generally because of the large variety of industries. In most cases, effluents result from the following water uses:

- sanitary (washing, drinking, personal hygiene)
- cooling
- processing (manufacture, washing, waste and by-product removal, transportation)
- cleaning.

The detailed rate of discharge will vary from industry to industry and will depend significantly on the actual processes used. Industrial effluents can be highly variable (in both quantity and quality terms) due to such practices as batch discharges, operation start-ups and shut-downs and working hours distribution.

The processing liquors from main industrial processes tend to be relatively strong while wastewaters from rinsing, washing and condensing are comparatively weak. Industrial wastewaters often contain:

- extremes in organic content

- a deficiency of nutrients
- inhibiting chemicals (toxins, bactericides)
- organic compounds which are resistant to biodegradation
- heavy metals and accumulative persistent organics.

Constituents of industrial effluents affect the biodegradability of the wastewater that contain the waste and hence may inhibit conventional biological treatment. Indeed the concentration of pollutants that may be discharged is strictly controlled (see Table 4.5).

Table 4.5 *Effluent standards for discharges to sewers (Gledhill, 1986)*

Quality component	Range	Problem
pH	6-10	corrosion
suspended solids	200 mg/l low up to 500mg/l	sewer blockages, excess sludge
BOD	<500mg/l	wastewater treatment plant overload
FOGs	100 mg/l	fouling
inflammables / hydrocarbons	prohibited	hazardous, explosive vapours
temperature	max. 43°C	promotes corrosion, increases solubility
toxic metals	10 mg/l	causes treatment inhibition
sulphate	500-1000 mg/l	sewer corrosion, H_2S gas in sewer
cyanides	0-1 mg /l	treatment inhibition, HCN gas accumulation in sewer

The extreme variability and industry specific nature of these flows is amply illustrated in Table 4.6. This table lists water consumption, wastewater production and pollutant load for a comprehensive list of industries.

Most industrial premises will have a domestic component of their waste and ideally the estimation of this should be based on a detailed survey of facilities and their use. It is conventional to allow at least 50% of domestic *per capita* flows for each employee. Mann (1979) suggests that a figure of 40-80 litres *per capita* per eight-hour shift is appropriate. By comparison with Table 4.3 it can be seen that this represents about 20% of daily *per capita* output.

4.3 ESTIMATION OF INFILTRATION

The level of infiltration is site specific and is based on the following components (Stanley, 1975; Martin *et al.*, 1982):

- type of subsoil
- height of groundwater level above sewers
- standard of workmanship in laying pipes
- type of pipe joint, number of joints and pipe size
- total length of sewer (including house connections)
- number and size of manholes and inspection chambers
- age of system.

Table 4.6 *Industrial water usage and wastewater production (after Henze et al. 1995)*

Industry/ production	Water consumption	Specific waste-water production	Specific pollution volumes	Concentration in the effluent	Remarks
Dairies					t = tonne weighed in milk
Milk for consumption	0.7-2.0 m^3/t	0.7-1.7 m^3/t	0.4-1.8 kg BOD$_7$/t	500-1500 g BOD$_7$/t	Caution: pH-variations discharge/emission of
Cheese	0.7-3.0 m^3/t	0.7-2.0 m^3/t	0.4-2.0 kg BOD$_7$/t	1000-2000 g BOD$_7$/t	
Mixed production	0.7-2.5 m^3/t	0.7-2.0 m^3/t	0.7-2.0 kg BOD$_7$/t	1000-2000 g BOD$_7$/t	tp = tonne product
Slaughterhouse		3-8 m^3/tp		500-2000 g BOD$_7$/t	Caution: Strong smell, stiff hair, disinfectants
Slaughtering			7-16 kg BOD$_7$/t	10-20 g Tot-P/tp	Large variations in the water consumption depending on type of production
Slaughtering + meat specialities		3-12 m^3/tp	10-25 kg BOD$_7$/t	500-2000 g BOD$_7$/t	
		1-15 m^3/tp	6-15 kg BOD$_7$/t	500-1000 g BOD$_7$/t	
Meat specialities					m^3* = product
Breweries	3-7 m^3/ m^3*	3-7 m^3/ m^3			Caution: High pH
Beer and soft drinks			4-15 kg BOD$_7$/t	1000-3000 g BOD$_7$/m^3	t = ton raw material
Canneries					Caution: flotables
Potatoes (dry peel)	2-4 m^3/t		3-6 kg BOD$_7$/t	1000-2000 g BOD$_7$/m^3	
Potatoes (wet peel)	4-8 m^3/t		5-1kg BOD$_7$/t	2000-3000 g BOD$_7$/m^3	
Beetroots	5-10 m^3/t		20-40 kg BOD$_7$/t	3000-5000 g BOD$_7$/m^3	tf = ton finished product
Carrots	5-10 m^3/t		5-15 kg BOD$_7$/t	800-1500 g BOD$_7$/m^3	t = ton raw material
Peas	15-30 m^3/t		15-30 kg BOD$_7$/t	1000-2000 g BOD$_7$/m^3	
Mixed production (vegetables)	20-30 m^3/t			5000-10000 g BOD$_7$/m^3	
Fish	8-15 m^3/t	4-8 m^3/t	10-50 kg BOD$_7$/t		t = ton raw material
Textile industry					
The whole industry	100-250 m^3/t	100-250 m^3/t	50-100 kg BOD$_7$/t	100-1000 g BOD$_7$/m^3	Caution: High water temp. extreme pH-value, chlorine gas, hydrogen sulfide gas, dangerous chemicals (allergies)
Cotton		100-250 m^3/t	70-120 kg BOD$_7$/t	200-600 g BOD$_7$/m^3	
Wool		50-100 m^3/t		500-1500 g BOD$_7$/m^3	
Synthetic fabrics		150-250 m^3/t	15-30 kg BOD$_7$/t	100-300 g BOD$_7$/m^3	
Tanneries			30-100 kg. BOD$_7$/t	1000-2000 g BOD$_7$/m^3	t = ton raw material
Tanneries			30-100 kg BOD$_7$/t	1000-2000 g BOD$_7$/m^3	t = ton raw material

Industry/ production	Water consumption	Specific waste-water production	Specific pollution volumes	Concentration in the effluent	Remarks
Mixed production	20-70 m³/t	20-70 m³/t	1-4 kg Cr/t	30-70 g Cr/m³	Caution: chromium, pH-variations, sludge and hair
Hides	20-40 m³/t	20-40 m³/t	0-100 kg S²/t	0-100 g S²/m³	
Fur	60-80 m³/t	60-80 m³/t	10-20 kg Tot-N/t	200-400 g Tot-N/m³	t = tonne of washing
Laundries	20-60 m³/t	20-60 m³/t	20-40 kg BOD₇/t	300-800 g BOD₇/m³	Laundries using counter-current wash have approx. 70% lower water consumption but the same emission of pollution (kg BOD₇/t)
Wet washing			10-20 kg Tot-P/t	10-50 g Tot-P/m³	Caution: High temp.
Galvanic industries	20-200 l/m²	20-200 l/m² <1 m³/h* max. 10m³/h	3-30 g hm/m² 2-20 CN/m²	Before own treatment approx. 150 g hm/m³ approx. 100 g CN/m³ After own treatment 1-10 g hm/m³ 0.1-0.5 g CN/m³	m² = m² surface area hm = heavy metals *50% of all galvano-industries have a flow < 1 m³/h Caution: Solvents, cyanide, extreme pH-value, heavy metals, complex builders
Electrical circuit industries	0.5-1.5 m³/m²	0.5-1.5 m³/m²	100-200 g Cu/m² 0-5g Sn/m² 0-5 g Pb/m²	100-200 g Cu/m³ 0-5g Sn/m³ 0-5 g Pb/m³	m² = m² laminate
Photolabs	0.5-1.5 m³/m²	0.5-1.5 m³/m²	200-400 g BOD₇/m²	400-700 BOD₇/m³ 50-100 g EDTA/m³	m² = m² emulsion There are large variations in the pollution Caution: Damage to the skin by contact, allergic reactions
Printing houses	30-40 m³/d	30-40 m³/d	approx. 7 kg Zn/d approx. 0.04 kg Ag/d approx. 0.03 kg Cr/d approx. 0.01 kg Cd/d	170-230 g Zn/m³ 1.0-1.3 g Ag/m³ 0.8-1.0 g Cr/m³ 0.2-0.3 g Cd/m³	The expenses are based on an investigation made in the trade. The table shows an average printer with a water consumption of 30-40 m³/d. Caution: Solvent, acids
Car repair/wash					
Cars	approx. 400 l/(Lt) approx. 200 l/(Ht)				Caution: Solvent Lt = low-pressure washing
Lorries	approx. 1200 l/(Ht)				Ht = high-pressure washing

Infiltration can only really be assessed by field measurements. This is complicated by the fact that infiltration levels can vary seasonally with catchment wetness. Several nominal rules have been proposed for making allowance for infiltration flows at the design stage and two of these are outlined below.

The amount of infiltration ranges widely from 0.01 m^3 to 1.0 m^3 per day per mm diameter per km length (Metcalf and Eddy, 1991). In old sewers, rates have been measured from 35 m^3 to 115 m^3 per day per km length. Infiltration along sewer lines usually arises as a result of poor design detailing or construction, and will generally increase as the system degrades physically.

In the UK, Stanley (1972) found rates in existing sewers subject to infiltration ranging from 15% to 49% of average DWF (19 to 102 litres *per capita* per day). Typical design guidance is to allow 10% of design *per capita* water consumption (in effect 20 litres/*capita* day), but in practice much higher figures are used. For smaller foul sewers serving housing estates, older editions of *Sewers for Adoption* suggested an allowance for infiltration of 10% of the design flow of 6 DWF.

4.4 DATA FOR WASTEWATER TREATMENT DESIGN

Schwinn and Dickson (1972) studied the magnitude, variability and inter-relationships between various pollutants at a treatment works (ammonia, total nitrogen, and phosphorus were measured in several raw domestic wastewaters).

The primary conclusions for the catchment studied were:

- for design purposes, ammonia concentrations can be assumed to be independent of the flow and of BOD and SS concentrations (nutrient loading is independent of flow)

- BOD and SS concentrations are inversely related to flow and positively correlated to each other, but the most extreme BOD and SS concentrations generally do not occur simultaneously

- no pronounced seasonal variations or differences between the days of the week in the concentrations data were found.

Aalderink's (1990) study analysed the correlation between dry weather flow rate and concentrations of pollutant. Out of six water quality parameters measured, there was **no** significant correlation between the flow rate and the constituent pollutant concentrations.

4.5 DATA FOR SEWERAGE SYSTEMS MODELLING

In computer models of sewer systems, diurnal profiles of flow can be specified. For in-sewer water quality modelling the three parameters usually required are the concentrations of suspended solids, BOD and ammonia. The Foundation for Water Research report on *Development of the Urban Pollution Database* (1994b) suggests the revised default values shown in Table 4.7 for use with the MOSQITO sewer modelling software.

Table 4.7 *Suggested default values for MOSQITO (after Foundation for Water Research, 1994)*

Parameter	Default value
Flow (litres/*capita* day)	210
Total suspended solids (mg/l)	250 = 52.5 g/*capita* day
BOD (mg/l)	210 = 44.1 g/*capita* day
COD (mg/l)	455 = 95.6 g/*capita* day
Ammonia (mg/l)	30 = 6.3 g/*capita* day

The other parameters quoted relate to surface sediments and pipe sediment characteristics. There does not seem to be any recommendation relating to DWF derived solids. No suggestions for the diurnal variation of flow or quality parameters are given.

5   Problems and data deficiencies

5.1   PROBLEMS WITH DWF DEFINITION

5.1.1   Delayed rainfall effects

The conceptual definition of the quantity of DWF is relatively straightforward, but there can be problems with interpretation. One of the main reasons for this is the inclusion of infiltration. Estimates of infiltration are difficult to arrive at theoretically and the only exact way of estimation for existing systems is by field measurement – but even this can lead to confusion.

Figure 5.1 shows a seven-day period following a period of rainfall. The last rain occurred about 09:00 on day 1 of the plot. Plotted against this are the flows from the following week during which no rain occurred. The graph indicates that indirect rainfall effects continue for up to seven days after the last rain before a constant diurnal pattern is achieved. This duration may vary with catchment type and antecedent catchment wetness.

To overcome this problem the Institute of Water Pollution Control (IWPC, 1975) defined dry weather flow as

The average daily flow to the treatment works during seven consecutive days without rain following seven days during which the rainfall did not exceed 0.25 mm on any one day.

The definition holds for wastewater that is essentially domestic in character. Where there is a significant industrial component it should be based on the flow on five working days. This is the most widely used in practice (IWEM, 1992)

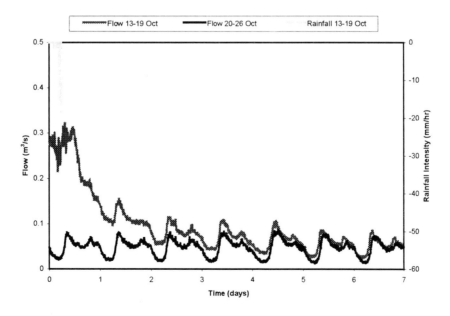

Figure 5.1 *Flow recession after rain*

5.1.2   Seasonal variation

There is, however, a significant difficulty with this definition due to seasonal variation.

Figure 5.2 shows the diurnal profiles for two dry weather days, in accordance with the above definition, for the same site in January and October. It can be seen that the two, whilst having a similar shape, are different in magnitude.

However, Figure 5.3 compares the two profiles after the removal of the infiltration component where the two can be seen to be almost identical. The initial difference would therefore seem to be due solely to a seasonal difference in the level of infiltration.

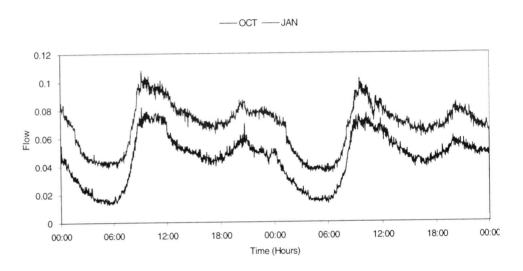

Figure 5.2 *Effects of infiltration*

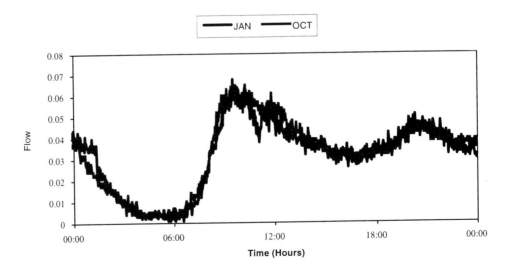

Figure 5.3 *Diurnal profiles after removal of infiltration*

An attempt to overcome this difficulty was made by the National Water Council (NWC, 1979), whose DWF definition states that it is

The median flow in dry weather, i.e. the median value over 24 hours of all days when rain did not exceed 1 mm (in four quarters of the year).

The median flow is that value which, when selected from a given number of all the eligible flows ranked in order of magnitude, forms the mid-point of the series. However it is still only calculated over a quarter-year period.

5.1.3 Delayed rainfall effects on quality

The above definitions consider only flow, but there are also the quality aspects to consider. It might be expected that once the effects of direct runoff have ended then the total load of BOD, ammonia and other quality parameters will return to those expected on other dry days.

Figure 5.4 shows the loads of suspended solids and BOD delivered to a treatment works site before, during and after rain. Rain occurred on day 3 and the other days were dry. On the dry days immediately after rain, the loads are higher than on the dry days before rain. One explanation for this is that the higher flows are scouring out sediment deposited on other dry days. The additional loads are unlikely to be derived from the rainfall. Further work is needed to determine whether this explanation is correct and whether the effect occurs on other catchments.

No definition of a dry day in terms of quality parameters has been found.

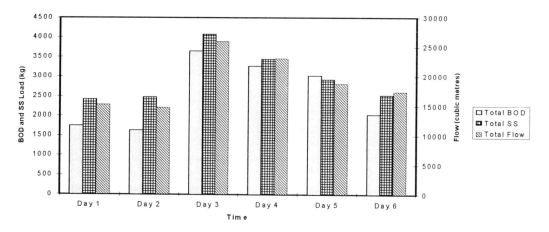

Figure 5.4 *Variation of loads to treatment*

5.2 DATA DEFICIENCIES

A summary of the needs of the various users of DWF was presented in Section 2.4. The currently used methodologies and figures in relation to DWF have been described in Sections 4.1 to 4.3. In this section a reconciliation is attempted to see how far current knowledge goes to meeting the needs. Significant shortfalls are identified.

The review of current practice has revealed a range of currently used values for domestic *per capita* outputs to sewers. Some of these values are intended to represent actual *per capita* outputs that may increase in the future or may already have increased since they were estimated. Other values are perceived to be design values and may allow for future increases in water consumption and may also allow for allowances for infiltration. It is also not clear if some of the values include for non-metered commercial inputs. Yet another area for confusion could also centre on the export and import of certain parameters, such as BOD, as people commute from home to work.

Guidance on how to estimate infiltration values appears to be limited.

Several references investigated attempts to characterise in-sewer quality parameters in terms of concentration, with a wide range of values quoted for each parameter. These references have not been quoted as it was stated in Chapter 3 that a more relevant measure for the major parameters was in terms of load. References quoting *per capita* outputs of the major parameters in terms of load were more limited and it seemed prudent to try to use available data to confirm the accuracy of the figures. However, there are some parameters that may impair the efficiency of treatment processes whose effect depends on concentration.

Some guidance is available on peak dry weather flow in sewers but virtually no guidance is available on the diurnal variation of dry weather flow and quality parameters. It is probably impractical to define default diurnal profiles for larger catchments due to their dependence on catchment characteristics. It may be possible, however, to derive default profiles which are attributable to small areas. This is discussed further in Chapter 6.

6 Analysis of data

6.1 INTRODUCTION

To characterise DWF, existing data has been analysed. A basic hypothesis was adopted for the analysis: that apparent differences in DWF characteristics between catchments are caused mainly by differences in infiltration rates; and that real differences are slight. Therefore, as far as possible, calculations were carried out using flows net of infiltration.

Data sets were obtained from 95 sites where at least some quality parameters were measured at frequent (hourly or shorter) intervals over a 24-hour period. The number of 24-hour data sets for each site ranged from 1 to 25. The data sets include those on the *Urban Pollution Database*; additional data supplied by water undertakings in England and Scotland; and daily summary data for 76 WwTW sites in England and Wales provided by the Smisson Foundation. Flow-only data was available for several more sites.

6.2 METHODOLOGY OF ANALYSIS

The analysis is concentrated on parameters for which data was available at hourly intervals over 24-hour periods: flow, SS, ammonia and either BOD or COD. There was insufficient information on other parameters, such as dissolved BOD, to justify detailed analysis.

As a first step, the data was expressed graphically by plotting each parameter against time. This facilitated the comparison of the diurnal profiles of flow, SS, ammonia and BOD/COD between the sites. Plotting all the data sets for each site on a single graph for each parameter enabled an assessment of the amount of scatter in the results between data sets to be made.

The concentration readings for the quality parameters were then converted to mass flow by multiplying by the flow rate:

concentration (mg/l) x flow rate (m^3/s) = mass flow (g/s) **(6.1)**

The total mass of each quality parameter over the 24-hour period was then calculated. This mass is equal to the area under the curve on a graph of mass flow against time. For the purposes of this study the area was calculated by applying the trapezoidal rule to the hourly data:

volume = (time step)/2* (first + last + (2 x intermediate)) **(6.2)**

The total volume of flow in the 24-hour period was also calculated. An average concentration over 24 hours for the quality parameters was then obtained by dividing the total mass by the total volume of flow. An example of these calculations is shown in Box 6.1.

Box 6.1 *Mass flow and flow volume calculation*

A	B	C	D	E	F
time	flow rate m³/s	flow volume m³	ammonia con-centration mg/l	ammonia mass flow g/s	ammonia mass g
00:00	0.035	118.8	34.9	1.222	3671
01:00	0.031	108.0	26.9	0.834	2495
02:00	0.029	102.6	19.3	0.560	1672
03:00	0.028	104.4	13.3	0.372	1342
04:00	0.030	111.6	12.4	0.372	1512
05:00	0.032	135.0	14.7	0.470	2417
06:00	0.043	208.8	21.1	0.907	6128
07:00	0.073	275.4	37.6	2.745	12985
08:00	0.080	268.2	56.7	4.536	13584
09:00	0.069	246.6	44.6	3.077	9100
10:00	0.068	225.0	29.2	1.986	6334
11:00	0.057	212.4	27.1	1.545	5565
12:00	0.061	212.4	25.3	1.543	5310
13:00	0.057	203.4	24.7	1.408	4536
14:00	0.056	199.8	19.9	1.114	4236
15:00	0.055	198.0	22.5	1.238	4267
16:00	0.055	203.4	20.6	1.133	5258
17:00	0.058	212.4	31.1	1.804	6138
18:00	0.060	217.8	26.7	1.602	5685
19:00	0.061	230.4	25.5	1.556	5230
20:00	0.067	235.8	19.9	1.333	4964
21:00	0.064	216.0	22.2	1.421	4147
22:00	0.056	185.4	16.2	0.907	4236
23:00	0.047	153.0	29.5	1.387	4927
00:00		68.4	34.9	1.326	
Total flow volume		**4584.6 m³**	**Total mass of ammonia**		**125737 g**

Average concentration = 125737/4584.6 = 27.4 mg/l

Columns A, B and D contain measured values. Columns C, E and F contain values calculated from Equations 6.1 and 6.2 as follows:

Cell C(I) = ((cell B(I) + cell B(I+1))/2) x (cell A(I+1) - cell A(I)) x 3600

Cell E(I) = cell D(I) x cell B(I)

Cell F(I) = ((cell E(I) + cell E(I+1))/2) x (cell A(I+1) - cell A(I)) x 3600

The total measured flow was then split into population-generated and infiltration components using the formula given in the CIRIA report *Control of Infiltration to Sewers (1997)*:

$$PG = \frac{(DWF - Min)}{F} \tag{6.3}$$

where

DWF	=	average daily dry weather flow
Min	=	minimum night-time flow
F	=	factor
PG	=	population-generated flow

One problem in using this equation is its sensitivity to the factor F, and this is discussed later in this Chapter. Examples of the calculation are shown in Box 6.2.

An average concentration of the quality parameters over 24 hours was also calculated by dividing the total mass by the total volume of population-generated flow.

Box 6.2 *Removal of infiltration and calculation of average concentration*

INFILTRATION = (average - minimum) / 0.9 = (0.53 - 0.28) / 0.9 = 0.025 m³/s
Columns A and B contain measured values
average flow = 0.053 m³/s
volume of infiltration in 24 hours = 0.025 x 24 x3600 = 2186 m³

A	B	C	A	B	C
time	flow m³/s	total flow - infiltration m³/s	time	flow m³/s	total flow - infiltration m³/s
00:00	0.035	0.010	13:00	0.057	0.032
01:00	0.031	0.006	14:00	0.056	0.031
02:00	0.029	0.004	15:00	0.055	0.030
03:00	0.028	0.003	16:00	0.055	0.030
04:00	0.030	0.005	17:00	0.058	0.033
05:00	0.032	0.007	18:00	0.060	0.035
06:00	0.043	0.018	19:00	0.061	0.036
07:00	0.073	0.048	20:00	0.067	0.042
08:00	0.080	0.055	21:00	0.064	0.039
09:00	0.069	0.044	22:00	0.056	0.031
10:00	0.068	0.043	23:00	0.047	0.022
11:00	0.057	0.032	00:00	0.038	0.013
12:00	0.061	0.036			

6.3 QUALIFICATION OF RESULTS

6.3.1 Data quality

Where data was at hourly intervals, values were frequently missing. If an isolated value was missing this was added by interpolation. If several data values were missing the data was rejected. The time of day at which measurements started varied between data sets. In order to standardise the data for comparison the data sets were rearranged to run from midnight to midnight.

Anomalies were noted in some of the flow data. There were sites where the measured minimum flow at night was less than that measured at a site upstream. This could indicate a bifurcation or significant exfiltration between the two sites, or an error in flow measurement at one or both of the sites.

At some sites the average measured flow varied significantly from day to day. Examination of the flow hydrographs suggested that in some cases these differences were probably due to measurement problems rather than real changes.

Little background information was available to confirm judgements on quality of flow data so all flow data has been taken at face value.

In some cases flow data was supplied as an integral part of the data set. In others the flow data was supplied separately from the quality data. There were indications with some of the separately supplied sets that the flow data had been collected to GMT, which is standard practice for flow monitoring, whereas the quality data had been collected to BST. For other sets the time base to which the clock times referred was not clear and an assumption had to be made. Because of the limited number of usable data sets for most sites, no differentiation has been made between weekday and weekend flows.

6.3.2 Estimation of infiltration

As mentioned earlier, the estimation of the infiltration component of measured flow is sensitive to the value used for F in Equation 6.3. In the most upstream parts of sewer systems F is thought to be in the range 0.88 to 0.92 in line with minimum night-time flows. Because of attenuation, this factor is likely to decrease in the downstream reaches of larger systems. F should also take into account flows from industrial premises, which may be working 24-hour days. As insufficient knowledge was available for some of the data sites the value of F to be used was uncertain. Experience in using the formula has also shown it to be very sensitive to the accuracy of the flow data – for average daily flows of less than 10 litres/s the flow rates need to be accurate to 0.1 litre/s.

In the absence of better information about the sites, a constant value of F = 0.9 has been used for all infiltration calculations. To illustrate the sensitivity to F, one of the data sets with the highest daily flow volume was considered. This location is described as being on a trunk sewer, and as such a lower value of F is likely to apply. For this site:

Average measured flow = 353 l/s
Minimum measured flow = 204 l/s

with F = 0.9, population generated flow = 166 litres/s
with F = 0.8 (allowing f-r attenuation at this downstream site), population-generated flow = 187 l/s.

6.4 *PER CAPITA* CONTRIBUTIONS

To establish the daily *per capita* contribution of flow, the total volume net of infiltration was correlated with the contributing population. The graph is presented in Figure 6.1, which also contains the best-fit regression line, its equation and correlation coefficient. The slope of the line corresponds to the *per capita* output of flow:

Net flow volume = (per capita output) x (population) **(6.4)**

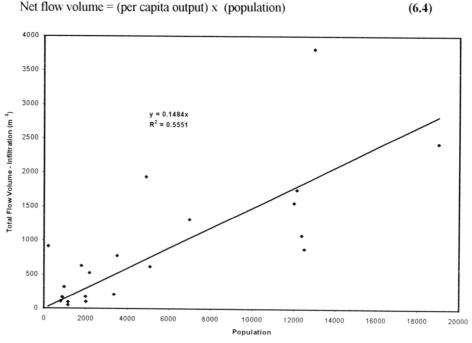

Figure 6.1 *Total flow volume minus infiltration against population*

Similar regressions were carried out for total daily loads of BOD, suspended solids and ammonia against population in order to estimate the daily *per capita* outputs of these parameters. The results of these regressions are presented in Table 6.1.

Table 6.1 *Calculated* per capita *flow and load*

	FLOW	**BOD**	**SS**	**NH4**
Per capita daily output from population (g)	148	67.7	56.0	6.2
Correlation coefficient	0.558	0.709	0.658	0.567

Unfortunately, the contributing populations were known for relatively few sites. This reduced data set, together with uncertainty about the reliability of some of the data, has resulted in a large amount of scatter as indicated by the correlation coefficient.

Because the above results were not felt to be conclusive it was decided to try to make as much use of the available data as possible by adopting an alternative methodology. The total daily loads for each quality parameter were now correlated against total daily flow volumes net of infiltration. The correlation of total mass of ammonia against net flow volume is presented in Figure 6.2.

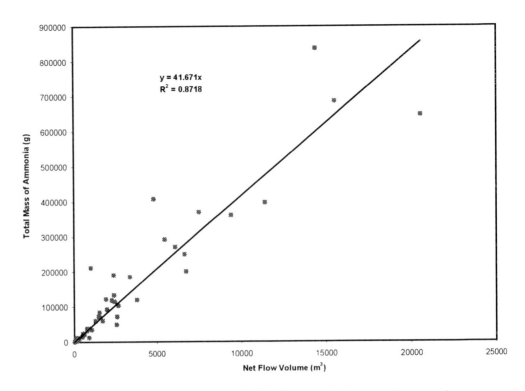

Figure 6.2 *Total flow volume minus infiltration against mass of ammonia*

The slope of this regression line now represents the concentration of ammonia net of infiltration.

Load = concentration x volume (6.5)

The correlation coefficient for this relationship, 0.87, was still low. Analysis showed it to be heavily influenced by the higher values on the right of the graph. Anomalies in the data are mentioned above. Using information about the catchment, three of the data points were recalculated. The points are indicated in Figure 6.2.

Point 1 is the point given as the example in the calculations showing the sensitivity to the factor F. This was replotted using the estimate of net flow calculated with F = 0.8. Point 2 is a site where the apparent infiltration was lower than at an upstream site on the same sewer. In this case the flow rates were adjusted to include additional infiltration and the masses of SS, BOD and ammonia were recalculated using the revised values. The third point is a site where sufficient background information was available to confirm the flow monitor was under reading. For this site the flow rates were corrected and the masses of the quality parameters recalculated.

The regression through the revised points (Figure 6.3) shows a much stronger correlation than was shown previously. Revised concentrations for suspended solids, BOD and ammonia and the new correlation coefficients are shown in Table 6.2.

However, the desired output of the analysis was not concentration but *per capita* load. In order to convert concentration to load, a figure for *per capita* flow of 140 litres/*capita* day was used.

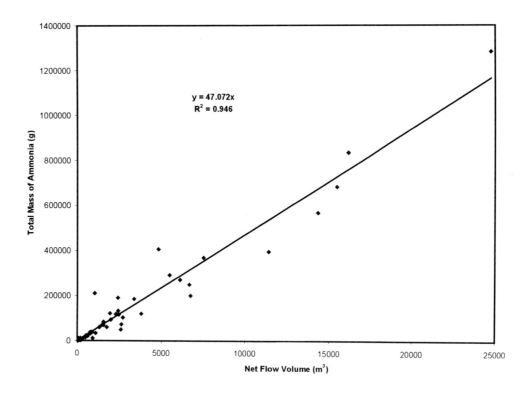

Figure 6.3 *Total flow volume minus infiltration against mass of ammonia*

Table 6.2 *Per capita loads from concentration*

	BOD	**SS**	**NH4**
Concentration based on net flow (mg/l)	383	392	47
Correlation coefficient based on net flow	0.857	0.926	0.946
Per capita daily output (g) (based on 140l/c.d)	53.6	54.9	6.58

It can be seen that these values of *per capita* output of BOD, SS and ammonia generally support those derived by the preferred method of analysis adopted earlier.

6.5   DIURNAL PROFILES

Examination of the diurnal plots illustrates some of the spatial and temporal variations referred to earlier. The effects of seasonally varying infiltration levels have already been discussed. The varying levels of infiltration also affect the shape of the diurnal curve expressed in dimensionless form by dividing through by the average flow. This is shown in Figure 6.4. Note that if infiltration is removed then the diurnal profiles for population-generated flow are identical.

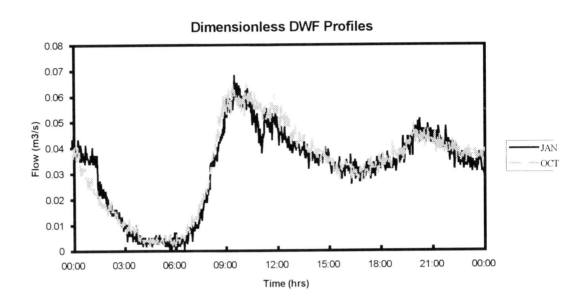

Figure 6.4   *Dimensionless DWF profiles*

Figure 6.5 shows diurnal profiles for the same day for two points along the same trunk sewer. From this it can be seen how the diurnal profile attenuates as it moves downstream. The amount of attenuation depends on the size and steepness of the sewer system.

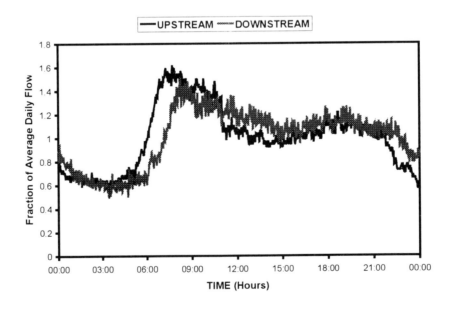

Figure 6.5   *Attenuation of diurnal profiles*

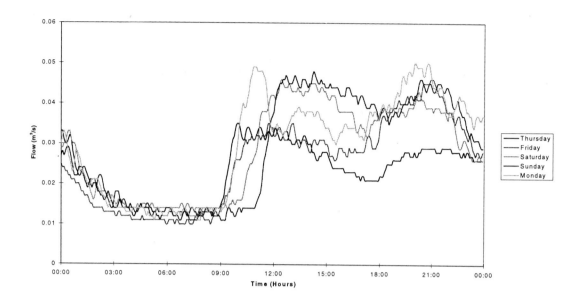

Figure 6.6   *Diurnal patterns over consecutive days*

Figure 6.6 compares the diurnal pattern of the inflow to a WwTW over five consecutive dry days. There are obvious differences, but it is interesting to note that the start of the morning increase in flow starts at the same time on all three working days, an hour later on the Saturday and an hour later still on the Sunday. Despite these differences there is very little difference in the peak flows.

Analysis of Figures 6.5 and 6.6 show the daily peaks barely exceeding 1.8 times the average flow. This might be partly because of the effects of infiltration. It is certainly the case that few of the dimensionless profiles examined exceeded a peak of twice the average flow.

The quality determinands also show a diurnal variation in concentration. Figure 6.7 shows the diurnal variation of ammonia at the upstream site used in Figure 6.5 over three dry days. It can be seen that a very consistent pattern is exhibited. Part, but not the whole, of this diurnal variation can perhaps be explained by the varying ratios of infiltration to population-generated flow through the day.

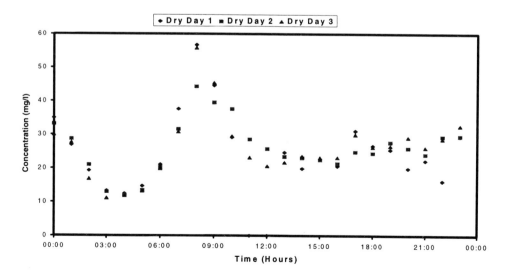

Figure 6.7 *Diurnal variation of ammonia concentration*

Like flow rate, this pattern tends to 'attenuate' in the downstream direction due to the processes of advection and dispersion in the flow. Figure 6.8 compares the upstream and downstream diurnal variations in ammonia concentration for the same sites used for flow in Figure 6.5.

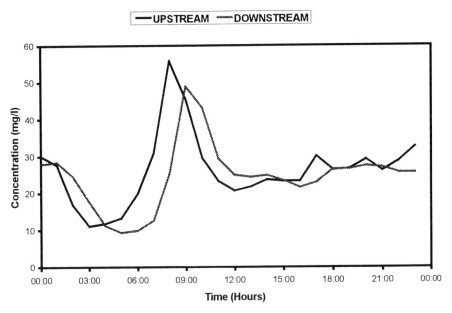

Figure 6.8 *Attenuation of diurnal ammonia profile*

For suspended solids and BOD such a clearly defined pattern is not apparent although there are consistent lower concentrations in the small hours of the morning. The patterns are shown in Figures 6.9 and 6.10.

Over short time scales ammonia is a largely conservative parameter, carried as a dissolved load. It is therefore easy to obtain representative samples and the analytical techniques are relatively robust. Suspended solids are more difficult to sample representatively and some BOD load is attached to suspended matter. The analytical results for these parameters are less consistent. Some of the observed variability is probably due to sampling and analysis errors.

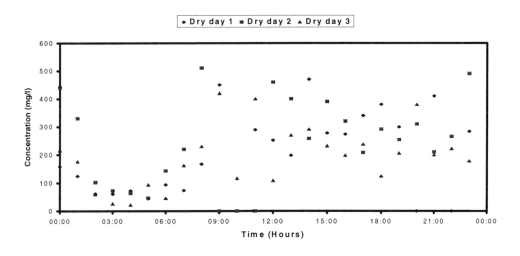

Figure 6.9 *Diurnal variation of suspended solids concentration*

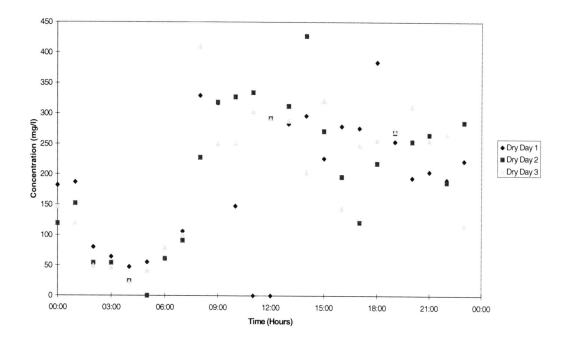

Figure 6.10 *Diurnal variation of BOD concentration*

6.6 SEASONAL VARIATION

Very few of the data sets contained a range of events that covered a long time period, so it has not been possible to review long-term effects – such as changing infiltration levels – on the various parameters.

6.7 MODELLING REQUIREMENTS

Per capita inputs, which are one of the requirements for modelling for both in-sewer flows and quality and WwTW flows and quality, have already been discussed in Section 6.4 above. Another requirement is the diurnal variation of the flow and quality parameters.

For WwTW modelling it is not believed possible to arrive at default diurnal profiles as, due to the effects of attenuation and advection and dispersion, they are dependent on the size, flows and general nature of the catchment. However, profiles (and loads) can be generated from a suitable in-sewer model.

The diurnal profiles to be applied to in-sewer models should be appropriate to the size of the individual contributing areas, and this is especially necessary for the quality parameters. As with WwTW modelling, defining default diurnal profiles may not be possible in terms of total flow because of the effects of varying infiltration levels. It is therefore suggested that population-generated flow and infiltration are input as two separate entities into the model. This would permit the defining of default diurnal profiles. It would also allow default values of parameters based on flows net of infiltration to be defined.

Most detailed models of sewerage systems typically divide the catchment into contributing areas in the range 1 ha to 5 ha. Default diurnal profiles should therefore be developed based on areas of about this size. A difficulty with doing this is that the flows generated are so small that they can be difficult to measure. To illustrate the principles an attempt to define diurnal profiles for one site, X, which has one of the smallest daily total flows, is detailed below.

Figure 6.11 shows a typical daily flow hydrograph for the site. It can be seen that the hydrograph is stepped as flows are only presented at a resolution of one litre/second. This lack of resolution affects the accuracy with which Equation 6.3 for separating infiltration can be used. There is a consequent knock on effect in defining a default diurnal profile for flow that is net of infiltration.

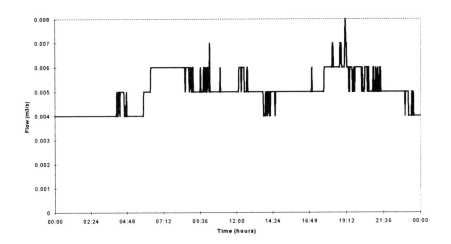

Figure 6.11 *Total measured daily flow at Site X*

It has been noted that measurements of ammonia concentration tend to be fairly consistent across most of the data sites. If a definitive value for net ammonia concentration could be arrived at then this may provide a better means of separating infiltration. The methodology would be to find a value of infiltration which, when applied in the way illustrated in the examples given in Section 6.4, produced an average adjusted ammonia concentration equal to the accepted default net concentration. This methodology has been carried out for one event for Site X assuming a default net ammonia concentration of 47 mg/l in line with the calculations in Section 6.4. The resulting value of infiltration was used to calculate adjusted concentrations for ammonia and SS at hourly intervals. These adjusted values were then converted to dimensionless profiles by dividing through by the average net concentrations.

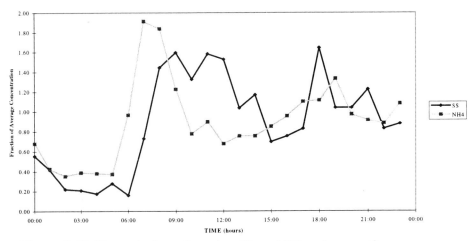

Figure 6.12 *Dimensionless diurnal profiles of SS and ammonia*

From Figure 6.12 it can be seen that ammonia and suspended solids concentrations have different diurnal profiles and different default profiles should therefore be specified. These profiles have been based on a single event for a single site. Default profiles should ideally be developed from a range of events for a range of sites. It is believed that there are not enough appropriate sites with suitable data in the current data set to complete this exercise.

Another aspect that it was hoped would be addressed from a modelling point of view was the settleability of the suspended sediments in the DWF and their effect on the measured concentrations of SS and BOD or COD. Again little data was available with which to attempt this. Most available analyses seem to have been carried out on sediment samples rather than samples collected from the flow.

6.8 CONCLUSIONS FROM THE ANALYSIS

6.8.1 The basic hypothesis

The basic hypothesis behind this analysis is that apparent differences in DWF characteristics between catchments are caused mainly by infiltration. To demonstrate this, the effects of infiltration have been removed.

Much of the analysis supports the hypothesis, but lack of background knowledge of the study catchments and uncertainties over the accuracy of the data makes it difficult to be sure. The hypothesis, and some of the methodologies for analysing DWF developed as part of this project, should be used as a framework for further analysis leading to a greater understanding of the nature of DWF.

6.8.2 Per capita outputs

Absence of population data for a significant proportion of the data sets has made it difficult to establish definitive values for *per capita* outputs of flow and the various quality parameters. However, values established using the limited data sets with population available have been supported by an alternative technique. This alternative technique also established concentrations for BOD, SS and ammonia that are net of infiltration.

Inaccurate flow rate measurement tends on the whole to lead to under-estimation rather than over-estimation of infiltration rates. Furthermore, under-estimation of flow will lead to under-calculation of the masses of the quality parameters. As no data sets with complete data have been rejected the inclusion of possibly erroneous data could lead to a bias in the regression analysis. There is therefore a probability that the derived *per capita* and net concentrations are too low.

6.8.3 Diurnal variation

As with other parameters it was hoped to be able to develop some standard diurnal profiles for flow and the major quality parameters by removing the effects of infiltration. The results of the analysis have shown that the dimensionless hydrographs can vary from location to location even within the same sewerage system. These differences have been attributed to the processes of attenuation for flow and advection and dispersion for the quality parameters. There can also be temporal variations at the same site with differences between weekday and weekend profiles. In principle it should be possible to develop a series of profiles for catchments of different sizes, but even this might not prove satisfactory by itself as these may all need adjustment for differences in topography.

An initial set of default diurnal profiles, appropriate to relatively small areas (1-5 ha.), which would be of use in sewer system modelling has been developed. Three profiles are presented for flow, ammonia and suspended solids/BOD. They have been developed to represent an average condition and may need refinement for particular catchments. It has been demonstrated that even for a single site the diurnal patterns will vary from day to day and especially from weekday to weekend. The time shift from GMT to BST should also not be forgotten if measurements continue on a GMT basis during summer months.

6.8.4 Estimation of infiltration

The calculations to estimate infiltration in this report were carried out using Equation 6.3. Difficulties were encountered with the use of this formula. This was partly due to the requirement to estimate a value of the factor F and partly due to not being able to make a sufficiently accurate estimate of the minimum flow at sites where the average flow was low and flow rate was available only in litre/second increments.

If, as appears to be the case, ammonia loads *per capita* for domestic wastewater are relatively constant across the range of catchment areas, then ammonia concentrations may offer an alternative methodology for estimating infiltration. It appears that ammonia can be measured reasonably accurately and consistently. Infiltration, therefore, could be estimated by finding the flow rate necessary to adjust the measured ammonia concentration to the expected net average value using the calculations set out in Box 6.2.

Overall infiltration in the analyses for this study ranged from 0% to 89% of total measured flow, with a mean of 45%.

The catchment with the highest calculated infiltration served a relatively small population and was therefore one to which the default value of F = 0.9 would be likely to apply. The quality parameter concentrations for this site were also very low, tending to confirm the analysis of very high infiltration.

6.8.5 Sampling and analysis

The quality results for ammonia for most sites analysed produce reasonably consistent results from day to day, whereas the results for suspended solids and BOD have contained a large amount of scatter. It is believed that in part this scatter could be due to prior deposition of suspended solids and associated BOD upstream in the sewer system and in part, to inconsistencies in the sampling and analysis techniques for these two variables. This has implications for WwTW operators who are trying to demonstrate compliance with percentage removal effluent consents, where crude wastewater load and concentration is the starting point, as in the UWWTD requirement for 50% removal of solids during the primary sedimentation process.

At the moment compliance will be judged against samples taken at the works inlet and at the outlet from the primary tanks where there is the probability that both samples could be in error. There are probably larger sources of error in the crude wastewater rather than the more homogeneous settled wastewater. If some of the solids load is transported by bedload in the sewer system it may not be picked up in the sampling for suspended solids thus leading to an under-estimation of the solids load removed. There could also be problems with sewer systems that deposit sediment in sewer during dry weather only for the deposited load to be flushed out during storms. Because of in-sewer deposition the suspended solids measurements at the inlet may be untypically low in dry weather. However, the deposited material which is scoured out during storms should prove readily settleable and would then be removed in the primary tanks, even at high inflow rates. Over the period of a year 50% removal of in-sewer solids may well be achieved.

The overall effect of these factors is likely to be an under-estimate of actual incoming wastewater loads in many catchments and hence an under-estimate of suspended solids removal against consent targets. In cases where meeting 50% removal may otherwise involve adding chemical dosing and extra sludge treatment, better analysis methods are advisable.

If firm values for the DWF parameters could be established they could be used as inputs to in-sewer water quality models. These models could predict the likely behaviour of the sewer system and develop theoretical inlet concentrations for suspended solids against which the measured settled wastewater concentrations could be judged.

The values of parameters required would include information on the properties of the settleable solids so that their in-sewer behaviour could be predicted accurately. Such properties would also be of interest to WwTW designers and modellers so that percentage removals in primary tanks could be estimated more accurately. In this context the ratio of volatile to total suspended solids is also required.

During the development of the UPM procedures solids analysis seems to have concentrated on sediment sampling rather than suspended solids sampling for settleable solids, particle size and other properties. Further work on this aspect is required.

7   Guidance for future practice

7.1   INTRODUCTION

This chapter gives new guidance, where the study results, even if of an interim nature, warrant it and provide a useful step forward.

Where no new guidance is given the reader should refer to the tables of typical practice in Chapter 4 or, preferably, the background references direct.

Users of hydraulic and water quality analysis software are recommended to check any default values of data used to see if these should be amended to the guidance given here.

Guidance is given on both parameter values and on methodology.

7.2   *PER CAPITA* CONTRIBUTIONS

The analysis carried out here supports the view that the flow and load domestic contributions shown in Table 7.1 are valid default parameter values for UK catchments.

Table 7.1 *Parameter values for UK catchments*

Parameter	Contribution
Wastewater flow (litres/*capita* day)	140-150
SS (g/*capita* day)	55-60
BOD$_5$ (g/*capita* day)	55-60
Ammonia (g/*capita* day)	6-7

Values towards the high end of the quoted ranges may be valid for smaller, steeper catchments with very little in-sewer deposition of load, or where there is more unmeasured commercial contribution counted in with the domestic. Lower values may be valid for larger, longer travel time catchments which are more prone to in-sewer deposition, or in catchments with very little non-domestic contribution included in the domestic DWF component, PG. For Scotland the *per capita* wastewater flow figure should possibly be increased by 25% in line with the statement in Section 4.1.1.

Parameter values outside these ranges, derived from crude wastewater sampling and analysis, particularly lower ones, should be viewed with suspicion, and not accepted as better than 'default' values without rigorous review of sampling and analysis procedure (see Section 7.5) and catchment specific reasons for the difference.

The 220 litres/*capita* day figure (including infiltration) which is quoted as a default in BS8005:1987 should no longer be used.

7.3 DIURNAL VARIATION

For key studies and major catchments the following interim dimensionless profiles are suggested for modelled areas of 2 ha to 5 ha. These are presented in Table 7.2.

Table 7.2 *Dimensionless profiles for DWF parameters*

Time	Flow	BOD/SS	NH4
01:00	0.25	0.6	0.87
02:00	0.08	0.6	0.87
03:00	0.07	0.6	0.87
04:00	0.08	0.6	0.87
05:00	0.11	0.6	0.87
06:00	0.58	0.6	0.87
07:00	1.63	0.9	1.30
08:00	2.12	1.2	1.74
09:00	1.83	1.2	1.30
10:00	1.68	1.2	0.87
11:00	1.51	1.2	0.87
12:00	1.23	1.2	0.87
13:00	1.14	1.2	0.87
14:00	1.12	1.2	0.87
15:00	0.92	1.2	0.87
16:00	0.95	1.2	0.87
17:00	1.16	1.2	0.87
18:00	1.26	1.2	0.87
19:00	1.44	0.65	0.87
20:00	1.16	0.60	0.87
21:00	1.05	0.75	0.87
22:00	1.11	0.66	0.87
23:00	0.96	0.73	0.87
24:00	0.55	0.60	0.87

The diurnal flow hydrograph is tabulated in dimensionless form. The values for BOD/SS and ammonia are tabulated as fractions of an average concentration net of infiltration. When combined with flows generated at 140 litres/*capita* day and using average concentrations from Table 6.2 (390 mg/l for BOD/SS and 47 mg/l for ammonia) they will produce daily *per capita* loads in line with Table 7.1.

Profiles for larger areas can then be developed by the use of the computer model of the sewerage system. The profiles have been developed net of infiltration and the infiltration component of DWF has to be added as a separate entity. Over the time scale of 24 hours it could reasonably assumed to be constant, but over the time scale of a year or more it may vary considerably (see Section 7.7). The assumption is also that infiltration does not contribute to the quality parameters of BOD, SS or ammonia, but it could contribute to chloride and sulphate concentrations due to ingress of infiltration water containing these, particularly seawater.

For new residential developments, the peak flows quoted in BS8005:1987 and in *Sewers for Adoption* are based on high litres/*capita* day, but are probably safe given that sewer sizes are determined by nearest available larger pipe size rather than actual peak flow rate.

For larger catchments, where sewer modelling is not being used to generate the profile, the US practice (Figure 4.2) is recommended, as it is based on the most real data. However peak flow attenuation is really related to storage and length in the sewer system, which will often be larger, for the same population, in the lower-density US areas than in the UK. Therefore the more extreme lines are safer to use.

In any case, note that this data indicates that peak flow will be lower than the old standard '3 x DWF' at any population larger than 40,000.

7.4   INFILTRATION

Infiltration should be estimated from flow survey data using the method proposed in the CIRIA report *Control of infiltration to sewers*.

This makes difficult demands in estimating the factor 'F' for larger catchments and in having sufficiently accurate and reliable night-time low-flow measurements in small catchments. Most standard temporary flow monitors used in the UK cannot provide accurate enough low-flow data for this analysis.

Where the results of the method are uncertain, the results of this study indicate that a check against the ammonia load using 47 mg/l and concentration (Box 6.2) can be used as a check or as a preferred estimate when the minimum-flow data is known to be unreliable.

The best estimate for long-term infiltration is from existing local catchments with similar conditions. Data is best analysed in terms of litres/day/km/mm of diameter before transfer and application.

In the absence of any other data for new sewerage systems in areas with high groundwater levels it is suggested that an allowance based on the average finding in Section 6.8.4, i.e. 45% of total DWF (including infiltration), is made. For purely domestic catchments this would mean calculating infiltration based on 115 to 120 litres/*capita* day. This compares to the 20 litres/*capita* day allowed in BS8005 and 120 litres/*capita* day effectively allowed in *Sewers for Adoption* (Water Services Association, 1995).

Infiltration dilutes wastewater. At the site with the highest infiltration (in the data analysed in Chapter 6), the wastewater concentrations were so low that they would probably be lower than the consent standards of some WwTW effluents. There may therefore be a case for not including infiltration in the DWF term in Formula A.

7.5 ESTIMATING CRUDE WASTEWATER LOAD FOR UWWTD 50% REMOVAL IN PRIMARY SEDIMENTATION

The low estimates of wastewater load that can arise from dry weather deposition in the sewer (and sampling and analysis difficulties) may give misleading results for suspended solids removal in primary sedimentation. These may lead to unnecessary investment and operational cost in chemically assisted sedimentation and additional sludge treatment and disposal expenditure.

When such estimates are being made they should be assessed using three separate methods:

1. A year's crude wastewater sampling and analysis, covering both dry weather and all storm flow periods and succeeding days (when high previously deposited loads may be washed through and settled out).

2. A 'source load' calculation using values given in Section 7.2.

3. A mass balance comparison against the combination of annual settled wastewater suspended solids load plus annual dry solids primary sludge make (after allowing for any co-settled surplus activated sludge).

All of these should be compared and reconciled as far as possible before drawing conclusions on percentage suspended solids removal in primary sedimentation.

7.6 DATA COLLECTION

One of the points which has come to light during the analysis of existing data was the amount of water quality data which had its value reduced by apparently inaccurate flow rate measurement, or failure of the flow monitor. Collecting and analysing water quality samples from sewers is an expensive exercise. It would therefore seem profligate not to get full value from it by ensuring accurate and consistent flow rate measurement.

Sewer flows from small areas tend to be correspondingly small and therefore difficult to measure accurately with conventional sewer flow monitors. The correspondingly small depths of flow, especially during the early hours of the morning, also make the collection of representative samples difficult. The choice of an appropriate method of flow measurement to achieve the desired accuracy and reliability is therefore crucial.

In order to achieve better flow measurement one possible method would be to install a temporary flume in a manhole with the upstream depth being measured by a downward pointing ultrasonic monitor. It has been argued that the installation of such a flume may represent an obstruction to flow, which might cause the upstream sewer to fail its level-of-service requirements under storm conditions. However, such a temporary is known to have been installed in a foul sewer of 150 mm diameter without apparent problem when a conventional ultrasonic Doppler flow monitor located in a similar manhole was in danger of causing blockage because of ragging. Flumes do not represent an obstruction to flow that might trap suspended particles or bedload. Flumes for insertion in sewer pipes of various sizes are manufactured in the USA where revenue generation is based on the amount of wastewater transported and treated. The presence of a flume would also create a larger upstream depth, which might facilitate the collection of quality samples from small areas. On the negative side, the reduced upstream velocities might give rise to deposition of sediment and hence no representative samples. Also, samples should ideally be taken just downstream of an area of high turbulence, which ensures mixing of the flow.

7.7   DEFINITION OF DWF

One of the main concerns expressed by consultees with the current definition of DWF was the inclusion of, and variability of, infiltration. Infiltration represents a significant unknown, which can only really be quantified by field measurement. It also varies seasonally so the question arises as to which value to use in the calculation of DWF.

To try to deal with this problem, efforts have been made here to characterise DWF in sewers in the absence of infiltration. Infiltration has therefore to be added as a separate entity at a later stage. Even without infiltration there are catchments where the DWF can vary with the time of year. In holiday resorts for example the summer population can be significantly greater than the winter one, and the *per capita* water usage and wastewater contribution may be different.

At least three possible DWF values could exist for a catchment, and each has its appropriate use:

1. Annual average DWF – based on annual average population, generated flow and infiltration – used for estimating long-term running costs of pumping stations and treatment works, and sludge treatment and disposal.

2. Winter DWF – a worst case based on winter population-generated flows and highest infiltration – used for assessing possible maximum flows for sewer design and maximum daily dry-weather peak flow to treatment.

3. Summer DWF – a worst case based on summer population-generated flows and low infiltration – used for setting WwTW effluent standards or modelling of river impact (including maximum consented loads) against low flows and high temperatures; and in estimating worst (lowest velocity) sewer conditions for deposition, odour and hydrogen sulphide generation.

8 Recommendations for further work

8.1 REFINEMENT OF CHARACTERISATION

One of the problems encountered with trying to characterise DWF with the existing data sets was lack of background information about the areas contributing to the measurement sites.

In order to refine the characterisation of domestic flows, to relate them better to catchment characteristics, it will be necessary to collect a new data set to ensure that data to cover all the aspects raised in this report are measured. The areas sampled must contain only domestic properties. Selected sites should be distributed throughout the UK and give a representative coverage of the topographical types; steep, moderate and flat. In order to meet the requirement of uniform development the areas selected might necessarily have to be small. Data from small areas could be used to establish the required default diurnal profiles for sewerage modelling purposes.

Twenty-four-hour quality sample sets should be taken at regular intervals over the period of at least a year so that sampling at different levels of infiltration is included and seasonal variations can be examined. Flow measurements should be continuous throughout the period. Dry days should be selected in line with the varying current definitions, i.e. after 24 hours with less than 1 mm of rain, after seven dry days, etc., in order to assess the effect of these definitions on quality parameters.

Data to be collected should include parameters to assess the settlement properties of the suspended matter in the flow.

If a new data collection exercise is carried out then some of the parameters of dry weather flow which are currently not of high general interest could be included in the analysis. Some of these parameters, in particular nutrients, microbiological parameters, and oestrogen mimics could become issues in the future. It would be prudent to investigate some of these further.

8.2 SAMPLING AND LABORATORY PROCEDURES

Further research would also appear to be needed into the sampling and laboratory procedures for suspended solids and BOD. The review of existing data has shown a large variation in these parameters from day to day. Better indication of how much of this variance is due to sampling, how much to laboratory procedures, and how much due to real changes is necessary to allow better characterisation of these parameters. Alternatively, new methods should be sought which minimise this variance. It is believed that the analytical procedures for measuring these parameters are relatively robust but care and attention needs to be taken in the laboratory.

8.3 OTHER ISSUES

Several of the consultees cited fats, oils and grease as an area of concern about which there appeared to be little information. The literature search for this project did not turn up any significant details on this topic although it has been the subject of an earlier CIRIA report.

Aesthetic pollutants, especially flotables, were another issue raised by some consultees. It is not clear what aspects of aesthetic pollutants should be investigated. It would appear that the primary concerns are the removal of such pollutants at scum boards and screens at overflows and at screens at the inlets to WwTW. Research is continuing in the UK in this area (Ruff *et al.* 1993 and Balmforth *et al.*, 1996).

Some research into the sources and transport of aesthetic pollutants is currently in progress (Butler and Graham *et al.* 1995). One method of reducing the impact of aesthetic pollutants may be to educate the public not to dispose in the WC substances that might cause aesthetic pollution (Souter *et al.*, 1996).

A final area in need of further investigation is in-sewer processes that may cause changes to the quality parameters during their transport through the sewer system (Hvitved-Jacobsen *et al.*, 1995).

Bibliography

AALDERINK, R H (1990)
Estimation of storm water quality flow characteristics and overflow loads from
treatment plant influent data
Wat. Sci. Tech., 22, 10/11, 77-85

ASCE/WPCF (1982)
Manual of Practice for Gravity Sanitary Sewer Design and Construction

ANDOH, R Y G and SMISSON, R P M (1996)
The practical use of wastewater characterisation in design
Wat. Sci. Tech., 33, 9, 127-134

ARUNDEL, J (1995)
Sewage and industrial effluent treatment. A practical guide
Blackwell Science, Oxford

ASHLEY, R M and DABROWSKI, W (1995)
Dry and storm weather transport of coliforms and faecal streptococci in combined
sewage
Wat. Sci. Tech., 31, 7, 311-320

BALMFORTH D J, MEEDS, E and THOMPSON, B (1996)
Performance of screens in controlling aesthetic pollutants
(Hanover, Germany) *Proc. 7th Int. Conf. on Urban Storm Drainage*, 989-994

BARNES, D, BLISS, P J, GOULD, B W and VALLENTINE, H R (1981)
Water and wastewater engineering systems
Longman Scientific and Technical, Harlow

BARTLETT, R E (1979)
Public health engineering – Sewerage
Applied Science, London

BECKER, F A, HEDGES, P D and SMISSON, R P M (1996)
The distribution of chemical constituents within the sewage settling velocity grading
curve
Wat. Sci. Tech., 33, 9, 143-146

BINNIE, C J A and HERRINGTON, P R (1992)
Possible effects of climate change on water resources and water demand
(ICE, London) *Symp. on Engineering in the Uncertainty of Climatic Change*

BRITISH STANDARDS INSTITUTION (1985)
Code of practice for Building Drainage
BS8301:1985

BRITISH STANDARDS INSTITUTION (1987)
Sewerage. Part 1 – Design 7 Construction
BS8005:1987

BOON, A G (1992)
Septicity in Sewers: causes, consequences, and containment
J. Instn. Wat. and Env. Mangt., 6, 2, 79-90

BURCHMORE, S and GREEN, M (1993)
Interim Report on Sewage-Derived Aesthetic Pollution
Foundation for Water Research, Report No. FR0339

BUTLER, D (1991)
A small scale study of wastewater discharges from domestic appliances
J. IWEM, 5, 178-185

BUTLER, D (1993)
The influence of dwelling occupancy and day of week on domestic appliance wastewater
discharges
Build. & Envir., 28, 1, 73-79

BUTLER, D and GRAHAM, N J D (1995)
Modelling dry weather wastewater flow in sewer networks
ASCE, J. Env. Engrng., 121, 2, Feb, 161-173

BUTLER, D, MAY, R W P and ACKERS, J C (1996)
Sediment transport in sewers Part 1: background
Proc. Instn. Civ. Engrs Wat., Marit. and Energy, 118, June, 103-112

CHARTERED INSTITUTE OF WATER AND ENVIRONMENTAL MANAGEMENT
(1996)
Water conservation and reuse
Proceedings of conference held in London

CHEBBO, G, MUSQUERE, P, MILISIC, V and BACHOC, A (1990)
Characterisation of solids transferred into sewer trunks during wet weather
Wat. Sci. Tech., 22, 10/11, 231-238

COMBER, S D W and GUNN, A M (1996)
Heavy metals entering sewage treatment works from domestic sources
J. CIWEM, 10, 2, 137-142

CONSTRUCTION INDUSTRY RESEARCH AND INFORMATION ASSOCIATION
(1997)
Control of infiltration to sewers

DAVIES, J W, BUTLER, D and YU, Y L (1996)
Gross solids movement in sewers: laboratory studies as a basis for a model
J. CIWEM, 10, 2, 52-58
London

DEGREMONT (1973)
Water Treatment Handbook, 4th edn.

DEPARTMENT OF ENVIRONMENT (1992)
Using water wisely. A consultation paper
DoE, London

EDWARDS, K and MARTIN, L (1995)
A methodology for surveying domestic water consumption
J. CIWEM, 9, 477-488

ESCRITT, L B (1959)
Sewerage and sewage disposal – calculations and design
Contractors Record, London

ESCRITT, L B (1984)
Sewerage and sewage treatment, international practice
Edited and revised by Haworth, W D. Wiley, Chichester

FAIR, G M, GEYER, J C and OKUN, D (1966)
Water and wastewater engineering, Vol 2: Water supply and wastewater removal
Wiley, Chichester

FIDDES, D and SIMMONDS, N (1981)
Infiltration – do we have to live with it?
Public Health Engr, 9, 1, 11-13

FOUNDATION FOR WATER RESEARCH (1994a)
Urban Pollution Management (UPM) – a planning guide for the management of urban wastewater discharges during wet weather
FWR FR/CL 0002 Manual

FOUNDATION FOR WATER RESEARCH (1994b)
The development of the urban pollution database
FR0441

FRIEDLER, E, BROWN, D M and BUTLER, D (1996)
A study of WC derived sewer solids
Wat. Sci. Tech., 33, 9, 17-24

GAINES, J B (1989)
Peak sewage flow rate: prediction and probability
J. Water Pollut. Control Fed., 61, 1241

GLEDHILL, P (1986)
Why investing in effluent treatment makes sense
Technology Ireland, July/August, 24-26

GRAY, N F (1989)
Biology of wastewater treatment
Oxford University Press, Oxford

HENZE, M, HARREMOES, JANSEN, and ARVIN (1995)
Wastewater treatment – biological and chemical treatment processes
Springer-Verlag

HVITVED-JACOBSEN, T, NEILSEN, P H, LARSEN, T and AAJENSEN, N (1995)
The sewer as a physical, chemical and biological reactor
Wat Sci. Tech., 31, 7, 328pp

IMHOFF, K, MULLER, W J and THISTLETHWAITE, D K B (1971)
Disposal of Sewage and Other Water-Borne Wastes, 2nd edn.
Butterworths

INSTITUTE OF WATER POLLUTION CONTROL (1975)
Glossary of terms in water pollution control
IWPC Manual of British Practice in Water Pollution Control

INSTITUTION OF WATER and ENVIRONMENTAL MANAGEMENT (1992)
Preliminary processes: a handbook of UK water practice
3rd edn.

JEFFERIES, C, YOUNG, H K and McGREGOR I (1990)
Microbiological aspects of sewage and sewage sludge in Dundee, Scotland
Wat. Sci. Tech., 22, 10/11, 27-52

LEDBURY, RW (1982)
Sewer system evaluation and rehabilitation
The Public Health Engr, 10, 4, 234-240

LUCK, B and ASHWORTH, D T (1996)
Maximum daily peak flow (MDPF) – a new approach to STW load prediction
Proc. Conf. on Sewer Modelling and Environmental Implications, Connemara, Ireland

MANN, H T (1979)
Septic tanks and small sewage treatment works
Report No. TR107, WRc

MARTIN, C, KING, D, QUICK, N J and NOTT, N A (1982)
Infiltration investigation, analysis and cost/benefit of remedial action, Paper No. 17,
ICE Conf. On Restoration of Sewerage Systems
Thomas Telford Ltd., London, 175-185

MEEDS, B and BALMFORTH, D J (1995)
Full-scale testing of mechanically raked bar screens
J. CIWEM, 9, Dec., 614-620

METCALF and EDDY, INC (1991)
Wastewater engineering: treatment, disposal and re-use; 3rd edn.
McGraw-Hill, New York

MINISTRY OF HOUSING and LOCAL GOVERNMENT (1970)
Final Report. Technical Committee on Storm Overflows and Disposal of Storm Sewage
HMSO, London.

NATIONAL BAG IT and BIN IT CAMPAIGN (1995)
Leedex Public Relations, London

NATIONAL RIVERS AUTHORITY (1995)
Saving water. The NRA's approach to water conservation and demand management.
NRA

NATIONAL WATER COUNCIL (1979)
Median flow in dry weather
NWC Bulletin, No. 43, Suppl. No. 117

NICOLL, E H (1988)
Small Water Pollution Control Works : Design and Practice
Ellis Horwood/Wiley, New York

OFWAT (1996)
Report on recent patterns of demand for water in England and Wales
Office of Water Services, May

ORHON, D (1994)
Modelling of activated sludge systems
Technomic

PAINTER, H A (1958)
Some characteristics of a domestic sewage
Water Pollution Research paper
Water and waste treat. j., 6, 496-498

PAYNE, J D and MOYS, G (1989)
Bacteria in urban drainage systems – literature review
Hydraulics Research Report, No. S.R 190, January 1989

PETTS, K W and STIFF, M J (1987)
An assessment of per capita loads of a domestic sewage – An interim report
WRc laboratory record 525-S

POMEROY, R D (1976)
The problem of hydrogen sulphide in sewers
Clay pipe development association

RUFF, S J, SAUL, A J, WALSH, M and GREEN, M J (1993)
Laboratory study of the gross particulate retention performances of large scale model
CSO structures
Proc. 6th Int. Conf. on Urban Storm Drainage, Niagara Falls, Canada, 1811-1815

RUMP, M E (1978)
The demand management of domestic water use
J. IWES, 33, 2, 173-182

RUSSAC, D A V, RUSHTON, K R and SIMPSON, R J (1991)
Insights into domestic demand from a metering trial
J. IWEM, 5, 342-351

SCHWINN, D E and DICKSON Jr, B H (1972)
Nitrogen and phosphorus variations in domestic wastewater
J. of Wat. Poll. Cont. Fed., 44, 11, 2059-2065

SHAW, V A (1963)
The development of contributory hydrographs for sanitary sewers and their use in sewer
design
Civ. Engr. S. Afr., 5, 9, 246

SIEGRIST, R, WITT, M and BOYLE, W (1976)
Characterisation of rural household wastewater
J. Env. Eng. ASCE, 102 (EE3), 533-548

SOUTER, N H, ASHLEY, R M, DAVIES, J W, BUTLER, D and STEWART, A (1996)
The UK bag-it-and-bin-it campaign: is it the most environmentally effective way of
dealing with domestic sanitary waste
Environmental Pollution – ICEP3, 249-256

STANLEY, G D (1975)
Design flows in foul sewerage sytems
DoE, London Project Report No.2

STEPHENSON, D and HINE, A E (1985)
Sewer flow models for various types of development in Johannesburg
IMIESA, October 1985

STEPHENSON, D and HINE, A E (1986)
Simulation of sewer flow
Municipal Engr., Instn. of Civ. Engrs., London, 107-112

TEBBUTT, T H Y (1992)
Principles of water quality control, 4th edn.
Pergamon Press

THISTLETHWAYTE, D K B (1972)
The control of sulphides in sewerage systems
Butterworths, Sydney

VERBANCK, M (1990)
Sewer sediment and its relation with the quality characteristics of combined sewer flows
Wat. Sci. Tech., 22, 10/11, 247-257

WATER SERVICES ASSOCIATION (1995)
Sewers for Adoption – a design and construction guide for developers, 4th edn.
WRc

WATER POLLUTION CONTROL FEDERATION/AMERICAN SOCIETY OF
CIVIL ENGINEERS (1969)
Design and construction of sanitary and storm sewers, Manual No.9
New York

WISE, A F E and SWAFFIELD, J A (1995)
Water, sanitary and waste services for buildings, 4th edn.
Longman Scientific and Technical, Harlow